McNicoll

MY INVENTED COUNTRY

MY INVENTED COUNTRY

*A Nostalgic Journey
Through Chile*

Isabel Allende

Translated from the Spanish
by Margaret Sayers Peden

HarperCollins*Publishers*

First published as *Mi País Inventado* in Spain in 2003 by Areté.

FIRST U.S. EDITION

Designed by Jennifer Ann Daddio

Map by David Cain

Printed on acid-free paper

Library of Congress Cataloging-in-Publication Data
Allende, Isabel.
[Mi país inventado. English]
My invented country: a nostalgic journey through Chile / by Isabel Allende; translated from the Spanish by Margaret Sayers Peden.—1st ed.
p. cm.
ISBN 0-06-054564-X
1. Allende, Isabel. 2. Chile—History—Coup d'état, 1973. 3. September 11 Terrorist Attacks, 2001. 4. Authors, Chilean—20th century—Biography. I. Peden, Margaret Sayers. II. Title.

PQ8098.1.L54 Z467   2003
863'.64—dc21   2002191267

03 04 05 06 07 NMSG/RRD 10 9 8 7 6 5 4 3 2 1

. . . for some reason or other, I am a sad exile.
In some way or other, our land travels with me
and with me too, though far, far away, live the
longitudinal essences of my country.

— PABLO NERUDA, 1972

Pacific
Ocean
BOLIVIA
Chuquicamata
Antofagasta
ATACAMA DESERT
LA SILLA
VALLEY OF ELQUI
ANDES MOUNTAINS
Valparaiso
Santiago
ISLAS JUAN FERNANDEZ
CHILE
ARGENTINA
Puerto Montt
CHILOE ISLANDS
Punta Arenas
TIERRA DEL FUEGO
PERU
Lima
BRAZIL
BOLIVIA
ISLA DE PASCUA (EASTER I.)
CHILE
Santiago
Buenos Aires
ARGENTINA
Punta Arenas
ANTARCTICA

A FEW WORDS OF
INTRODUCTION

Iwas born in the years of the smoke and carnage of the
Second World War, and the greatest part of my youth was
spent waiting for the planet to blow apart when someone dis-
tractedly pressed a button deploying atomic bombs. No one
expected to live very long; we rushed around swallowing up
every moment before being overtaken by the apocalypse, so
we didn't have time to examine our navels and take notes, as
people do today. In addition, I grew up in Santiago, Chile,
where any natural tendency toward self-contemplation is
quickly nipped in the bud. The saying that defines the lifestyle
of that city is "Shrimp that dozes is shrimp on the platter." In
other, more sophisticated cultures, like those of Buenos Aires
or New York, a visit to the psychologist was thought to be a
normal activity: to deprive oneself of that attention was considered evidence of a lack of culture or of mental deficiency.
In Chile, however, only dangerously disturbed patients visited
a psychologist, and then always in a straitjacket, but that
changed in the seventies, along with the arrival of the sexual
revolution. (One wonders if there's a connection . . .) In my
family no one ever resorted to therapy, even though many of
us were classic case studies, because the idea of confiding intimate matters to a stranger—and a stranger we were *paying* to
listen—was absurd. That's what priests and aunts were for. I

have very little training for reflection, but in recent weeks I have caught myself thinking about my past with a frequency that can only be explained as a sign of premature senility.

Two recent events have triggered this avalanche of memories. The first was a casual observation by my grandson Alejandro, who surprised me at the mirror scrutinizing the map of my wrinkles and said, with compassionate commiseration, "Don't worry, Grandmother, you're going to live at least three more years." I decided right then and there that the time had come to take another look at my life, in order to know how I wanted to live those three years that had been so generously granted.

The second event was a question asked by a stranger during a conference of travel writers where I'd been invited to give the opening address. I must make clear that I do not belong to that weird group of people who travel to remote places, survive the bacteria, and then publish books to convince the incautious to follow in their footsteps. Traveling demands a disproportionate effort, especially when it's to places where there is no room service. My ideal vacation consists of sitting in a chair beneath an umbrella on my patio, reading books of adventures I would never consider attempting unless I was escaping from something.

I come from the so-called Third World (what is the Second?), and I had to trap a husband in order to live legally in the First. I have no intention of going back to underdevelopment without good cause. Nevertheless, for reasons quite beyond my control, I have wandered across five continents, and have in addition been an exile and an immigrant. So I know something about travel, which is why I had been asked to speak at that conference. At the end of my brief talk, a hand was raised in the audience and a young man asked me what role nostalgia played in my novels. For a moment I was silent. Nostalgia . . . according to the dictionary, *nostalgia* is "a bittersweet longing for things, persons, or situations of the past. The condition of being homesick." The question took my breath away because until that instant I'd never realized that I write as a constant exercise in longing. I have been an outsider nearly all my life, a circumstance I accept because I have no alternative. Several times I have found it necessary to pull up stakes, sever all ties, and leave everything behind in order to begin life anew elsewhere; I have been a pilgrim along more roads than I care to remember. From saying good-bye so often my roots have dried up, and I have had to grow others, which, lacking a geography to sink into, have taken hold in my memory. But be careful! Minotaurs lie in wait in the labyrinths of memory.

Until only a short time ago, if someone had asked me where I'm from, I would have answered, without much thought, Nowhere; or, Latin America; or, maybe, In my heart I'm Chilean. Today, however, I say I'm an American, not simply because that's what my passport verifies, or because that word includes all of America from north to south, or because my husband, my son, my grandchildren, most of my friends, my books, and my home are in northern California; but because a terrorist attack destroyed the twin towers of the World Trade Center, and starting with that instant, many things have changed. We can't be neutral in moments of crisis. This tragedy has brought me face to face with my sense of identity. I realize today that I am one person in the multicolored population of North America, just as before I was Chilean. I no longer feel that I am an alien in the United States. When I watched the collapse of the towers, I had a sense of having lived a nearly identical nightmare. By a blood-chilling coincidence—historic karma—the commandeered airplanes struck their U.S. targets on a Tuesday, September 11, exactly the same day of the week and month—and at almost the same time in the morning—of the 1973 military coup in Chile, a terrorist act orchestrated by the CIA against a democracy. The images of burning buildings, smoke, flames, and panic are similar in both settings. That distant Tuesday in 1973 my life was split in two; nothing was ever again the same: I lost a country. That fateful Tuesday in 2001 was also a decisive moment; nothing will ever again be the same, and I gained a country.

Those two statements, the consoling words from my grandson and the question asked by a stranger at a conference, gave rise to this book. I'm not sure what direction it will take. For the moment, I'm wandering, but I ask you to stay with me a little longer.

I am writing these pages in a room perched high on a hill, under the vigil of a hundred gnarled oaks overlooking San Francisco Bay, but I come from a different place. Nostalgia is my vice. Nostalgia is a melancholy, and slightly saccharine, sentiment, like tenderness. It is nearly impossible to approach those emotions without sounding insipid, but I am going to try. If I fall and slip into cloying vulgarity I will climb out of it a few lines later. At my age—I'm at least as old as synthetic penicillin—you begin to remember things that have been erased from your mind for half a century. I haven't thought about my childhood or adolescence for decades. In truth, those periods of my remote past matter so little to me that when I look at my mother's photograph albums I don't recognize anyone except a bulldog with the improbable name of Pelvina López-Pun, and the only reason why she is etched in my mind is because we were very much alike. There is a snapshot of the two of us, when I was a few months old, in which my mother had to indicate with an arrow which of us was which. Surely my bad memory is due in part to the fact that those times were not particularly happy ones, but I suppose that's the case with most mortals. A happy childhood is a myth, and in order to understand that we have only to take a look at children's stories; for example, the one in which the wolf eats the beloved grandmother, then along comes a woodsman and slits the poor beast open with his knife, extracts the old woman, alive and uninjured, fills the wolf's belly with stones and then stitches him up, in the process creating such a thirst in the animal that he runs down to drink from the river, where he drowns from the weight of the stones. Why didn't they do away with him in a simpler, more humane way, is what I want to know. Surely because nothing is simple or humane in childhood. In those days there was no such term as "abused children," it was accepted that the best way to bring up little ones was with a strap in one hand and a cross in the other, just as it was taken for granted that a man had a right to give his wife a good shaking if his soup was cold when it reached the table. Before psychologists and authorities intervened, no one doubted the beneficial effects of a good switching. I wasn't whipped like my brothers, but I lived in fear, like all the other children I knew.

In my case, the natural unhappiness of childhood was aggravated by a mass of complexes so tangled that even today I can't list them. Fortunately, they left no wounds that time hasn't healed. Once I heard a famous Afro-American writer say that from the time she was a little girl she felt like a stranger in her family and her hometown. She added that nearly all writers have experienced that feeling, even if they have never left their native city. It's a condition inherent in that profession, she suggested; without the anxiety of feeling different, she wouldn't have been driven to write. Writing, when all is said and done, is an attempt to understand one's own circumstance and to clarify the confusion of existence, including insecurities that do not torment normal people, only chronic nonconformists, many of whom end up as writers after having failed in other undertakings. This theory lifted a burden from my shoulders. I am not a monster; there are others like me.

I never fit in anywhere: not into my family, my social class, or the religion fate bestowed on me. I didn't belong to the neighborhood gangs that rode their bikes in the street, my cousins didn't include me in their games, I was the least popular girl in my school, and for a long time I was the last to be invited to dance at parties—a torment, I like to think, due more to shyness than to looks. I cloaked myself in my pride, pretending it didn't matter to me, but I would have sold my soul to the devil to be part of a group had Satan presented me with such an attractive proposition. The source of my difficulties has always been the same: an inability to accept what to others seems natural, and an irresistible tendency to voice opinions no one wants to hear, a trait that frightened away more than one potential suitor (I don't want to give a false impression, there weren't very many). Later, during my years as a journalist, curiosity and boldness had their advantages. For the first time I was part of a community, I had absolute liberty to ask indiscreet questions and divulge my ideas, but that ended abruptly with the military coup of 1973, which unleashed uncontrollable forces. Overnight I became a foreigner in my own land, until finally I had to leave because I couldn't live and bring up my children in a country where terror reigned and where there was no place for dissidents like myself. During that period, curiosity and boldness were outlawed by decree. Outside Chile, I waited years to return once democracy was restored, but when that happened I didn't, because by then I was married to a North American and living near San Francisco. I haven't gone back to take up residence in Chile, where in truth I have spent less than half of my life, although I visit frequently. But in order to respond to the question that the stranger asked about nostalgia, I must refer almost exclusively to my years there. And to do that, I have to talk about my family because nation and tribe are confused in my mind.

MY INVENTED COUNTRY

COUNTRY OF LONGITUDINAL ESSENCES

L et's begin at the beginning, with Chile, that remote land that few people can locate on the map because it's as far as you can go without falling off the planet. *Why don't we sell Chile and buy something closer to Paris?* one of our intellectuals once asked. No one passes by casually, however lost he may be, although many visitors decide to stay forever, enamored of the land and the people. Chile lies at the end of all roads, a lance to the south of the south of America, four thousand three hundred kilometers of hills, valleys, lakes, and sea. This is how Neruda describes it in his impassioned poetry:

> *Night, snow and sand compose the form*
> *of my slender homeland,*
> *all silence is contained within its length,*
> *all foam issues from its seaswept beard,*
> *all coal fills it with mysterious kisses.*

This elongated country is like an island, separated on the north from the rest of the continent by the Atacama Desert—the driest in the world, its inhabitants like to say, although that must not be true, because in springtime parts of that lunar rubble tend to be covered with a mantle of flowers, like a wondrous painting by Monet. To the east rises the cordillera of the Andes, a formidable mass of rock and eternal snows, and to the west the abrupt coastline of the Pacific Ocean. Below, to the south, lie the solitudes of Antarctica. This nation of dramatic topography and diverse climates, studded with capricious obstacles and shaken by the sighs of hundreds of volcanoes, a geological miracle between the heights of the cordillera and the depths of the sea, is unified top to tail by the obstinate sense of nationhood of its inhabitants.

We Chileans still feel our bond with the soil, like the campesinos we once were. Most of us dream of owning a piece of land, if for nothing more than to plant a few worm-eaten heads of lettuce. Our most important newspaper, *El Mercurio,* publishes a weekly agricultural supplement that informs the public in general of the latest insignificant pest found on the potatoes or about the best forage for improving milk production. Its readers, who are planted in asphalt and concrete, read it voraciously, even though they have never seen a live cow.

In the broadest terms, it can be said that my long and narrow homeland can be broken up into four very different regions. The country is divided into provinces with beautiful names, but the military, who may have had difficulty memorizing them, added numbers for identification purposes. I refuse to use them because a nation of poets cannot have a map dotted with numbers, like some mathematical delirium. So let's talk about the four large regions, beginning with the *norte grande,* the "big north" that occupies a fourth of the country; inhospitable and rough, guarded by high mountains, it hides in its entrails an inexhaustible treasure of minerals.

I traveled to the north when I was a child, and I've never forgotten it, though a half-century has gone by since then. Later in my life I had the opportunity to cross the Atacama Desert a couple of times, and although those were extraordinary experiences, my first recollections are still the strongest. In my memory, Antofagasta, which in Quechua means "town of the great salt lands," is not the modern city of today but a miserable, out-of-date port that smelled like iodine and was dotted with fishing boats, gulls, and pelicans. In the nineteenth century it rose from the desert like a mirage, thanks to the industry producing nitrates, which for several decades were one of Chile's principal exports. Later, when synthetic nitrate was invented, the port was kept busy exporting copper, but as the nitrate companies began to close down, one after another, the pampa became strewn with ghost towns. Those two words—"ghost town"—gave wings to my imagination on that first trip.

I recall that my family and I, loaded with bundles, climbed onto a train that traveled at a turtle's pace through the inclement Atacama Desert toward Bolivia. Sun, baked rocks, kilometers and kilometers of ghostly solitudes, from time to time an abandoned cemetery, ruined buildings of adobe and wood. It was a dry heat where not even flies survived. Thirst was unquenchable. We drank water by the gallon, sucked oranges, and had a hard time defending ourselves from the dust, which crept into every cranny. Our lips were so chapped they bled, our ears hurt, we were dehydrated. At night a cold hard as glass fell over us, while the moon lighted the landscape with a blue splendor. Many years later I would return to the north of Chile to visit Chuquicamata, the largest open-pit copper mine in the world, an immense amphitheater where thousands of earth-colored men, working like ants, rip the mineral from stone. The train ascended to a height of more than four thousand meters and the temperature descended to the point where water froze in our glasses. We passed the silent salt mine of Uyuni, a white sea of salt where no bird flies, and others where we saw elegant flamingos. They were brush strokes of pink among salt crystals glittering like precious stones.

The so-called *norte chico,* or "little north," which some do not classify as an actual region, divides the dry north from the fertile central zone. Here lies the valley of Elqui, one of the spiritual centers of the Earth, said to be magical. The mysterious forces of Elqui attract pilgrims who come there to make contact with the cosmic energy of the universe, and many stay on to live in esoteric communities. Meditation, Eastern religions, gurus of various stripes, there's something of everything in Elqui. It's like a little corner of California. It is also from Elqui that our *pisco* comes, a liquor made from the muscatel grape: transparent, virtuous, and serene as the angelic force that emanates from the land. *Pisco* is the prime ingredient of the *pisco* sour, our sweet and treacherous national drink, which must be drunk with confidence, though the second glass has a kick that can floor the most valiant among us. We usurped the name of this liquor, without a moment's hesitation, from the city of Pisco, in Peru. If any wine with bubbles can be called champagne, even though the authentic libation comes only from Champagne, France, I suppose our *pisco,* too, can appropriate a name from another nation. The *norte chico* is also home to La Silla, one of the most important observatories in the world, because the air there is so clear that no star—either dead or yet to be born—escapes the eye of its gigantic telescope. Apropos of the observatory, someone who has worked there for three decades told me that the most renowned astronomers in the world wait years for their turn to scour the universe. I commented that it must be stupendous to work with scientists whose eyes are always on infinity and who live detached from earthly miseries, but he informed me that it is just the opposite: astronomers are as petty as poets. He says they fight over jam at breakfast. The human condition never fails to amaze.

The *valle central* is the most prosperous area of the country, a land of grapes and apples, where industries are clustered and a third of the population lives in the capital city. Santiago was founded in 1541 by Pedro de Valdivia. After walking for months through the dry north, it seemed to him that he'd reached the Garden of Eden. In Chile everything is centralized in the capital, despite the efforts of various governments that over the span of half a century have tried to distribute power among the provinces. If it doesn't happen in Santiago, it may as well not happen at all, although life in the rest of the country is a thousand times calmer and more pleasant.

The *zona sur,* the southern zone, begins at Puerto Montt, at 40 degrees latitude south, an enchanted region of forests, lakes, rivers, and volcanoes. Rain and more rain nourishes the tangled vegetation of the cool forests where our native trees rise tall, ancients of thousand-year growth now threatened by the timber industry. Moving south, the traveler crosses pampas lashed by furious winds, then the country strings out into a rosary of unpopulated islands and milky fogs, a labyrinth of fjords, islets, canals, and water on all sides. The last city on the continent is Punta Arenas, wind-bitten, harsh, and proud; a high, barren land of blizzards.

Chile owns a section of the little-explored Antarctic continent, a world of ice and solitude, of infinite white, where fables are born and men die: Chile ends at the South Pole. For a long time, no one assigned any value to Antarctica, but now we know how many mineral riches it shelters, in addition to being a paradise of marine life, so there is no country that doesn't have an eye on it. In the summertime, a cruise ship can visit there with relative ease, but the price of such a cruise is as the price of rubies, and for the present, only rich tourists and poor but determined ecologists can make the trip.

In 1888 Chile annexed the Isla de Pascua, mysterious Easter Island, *the navel of the world,* or Rapanui, as it is called in the natives' language. The island is lost in the immensity of the Pacific Ocean, 2,500 miles from continental Chile, more or less six hours by jet from Valparaíso or Tahiti. I am not sure why it belongs to us. In olden times, a ship captain planted a flag, and a slice of the planet became legally yours, regardless of whether that pleased its inhabitants, in this case peaceful Polynesians. This was the practice of European nations, and Chile could not lag behind. For the islanders, contact with South America was fatal. In the mid-nineteenth century, most of the male population was taken off to Peru to work as slaves in the guano deposits, while Chile shrugged its shoulders at the fate of its forgotten citizens. The treatment those poor men received was so bad that it caused an international protest in Europe, and, after a long diplomatic struggle, the last fifteen survivors were returned to their families. Those few went back infected with small pox, and within a brief time the illness exterminated eighty percent of the natives on the island. The fate of the remainder was not much better. Imported sheep ate the vegetation, turning the landscape into a barren husk of lava, and the negligence of the authorities—in this case the Chilean navy—drove the inhabitants into poverty. Only in the last two decades, tourism and the interest of the world scientific community have rescued Rapanui.

Scattered across the Easter Island are monumental statues of volcanic stone, some weighing more than twenty tons. These *moais* have intrigued experts for centuries. To sculpt them on the slopes of the volcanoes and then drag them across rough ground, to erect them on often-inaccessible bases and place hats of red stone atop them, was the task of titans. How was it done? There are no traces of an advanced civilization that can explain such prowess. Two different groups populated the island. According to legend, one of those groups, the Arikis, had supernatural mental powers, which they used to levitate the *moais* and transport them, floating effortlessly, to their altars on the steep slopes. What a tragedy that this technique has been lost to the world! In 1940, the Norwegian anthropologist Thor Heyerdahl built a balsa raft, which he christened *Kon Tiki,* and sailed from South America to Easter Island to prove that there had been contact between the Incas and the Easter Islanders.

I traveled to Easter Island in the summer of 1974, when there was only one flight a week and tourism was nearly nonexistent. Enchanted, I stayed three weeks longer than I had planned, and thus happened to be on the spot when the first television broadcast was celebrated with a visit by General Pinochet, who had led the military junta that had replaced Chile's democracy some months earlier. The television was received with more enthusiasm than the brand-new dictator. The general's stay was extremely colorful, but this isn't the time to go into those details. It's enough to say that a mischievous little cloud strategically hovered above his head every time he wanted to speak in public, leaving him wringing wet and limp as a dishrag. He had come with the idea of delivering property titles to the islanders, but no one was terribly interested in receiving them, since from the most ancient times everyone has known exactly what belongs to whom. They were afraid, and rightly so, that the only use for that piece of government paper would be to complicate their lives.

Chile also owns the island of Juan Fernández, where the Scots sailor Alexander Selkirk, the inspiration for Daniel Defoe's novel *Robinson Crusoe,* was set ashore by his captain in 1704. Selkirk lived on the island for more than four years—without a domesticated parrot or the company of a native named Friday, as portrayed in the novel—until he was rescued by another captain and returned to England, where his fate did not exactly improve. The determined tourist, after a bumpy flight in a small airplane or an interminable trip by boat, can visit the cave where the Scotsman survived by eating herbs and fish.

Being so far from everything gives us Chileans an insular mentality, and the majestic beauty of the land makes us take on airs. We believe we are the center of the world—in our view, Greenwich should have been set in Santiago—and we turn our backs on Latin America, always comparing ourselves instead to Europe. We are very self-centered: the rest of the universe exists only to consume our wines and produce soccer teams we can beat.

My advice to the visitor is not to question the marvels he hears about my country, its wine, and its women, because the foreigner is not allowed to criticize—for that we have more than fifteen million natives who do that all the time. If Marco Polo had descended on our coasts after thirty years of adventuring through Asia, the first thing he would have been told is that our *empanadas* are much more delicious than anything in the cuisine of the Celestial Empire. (Ah, that's another of our characteristics: we make statements without any basis, but in a tone of such certainty that no one doubts us.) I confess that I, too, suffer from that chilling chauvinism. The first time I visited San Francisco, and there before my eyes were those gentle golden hills, the majesty of forests, and the green mirror of the bay, my only comment was that it looked a lot like the coast of Chile. Later I learned that the sweetest fruit, the most delicate wines, and the finest fish are imported from Chile. Naturally.

To see my country with the heart, one must read Pablo Neruda, the national poet who in his verses immortalized the imposing landscapes, the aromas and dawns, the tenacious rain and dignified poverty, the stoicism and the hospitality, of Chile. That is the land of my nostalgia, the one I invoke in my solitude, the one that appears as a backdrop in so many of my stories, the one that comes to me in my dreams. There are other faces of Chile, of course: the materialistic and arrogant face, the face of the tiger that spends its life counting its stripes and cleaning its whiskers; another, depressed, crisscrossed by the brutal scars of the past; one that shows a smiling face to tourists and bankers; and the one that with resignation awaits the next geological or political cataclysm. Chile has a little of everything.

DULCE DE LECHE, ORGAN GRINDERS, AND GYPSIES

My family is from Santiago, but that doesn't explain my traumas, there are worse places under the sun. I grew up there, but now I scarcely recognize it, and get lost in its streets. The capital was founded following the classic pattern for Spanish cities of the time: a *plaza de armas* in the center, from which parallel and perpendicular streets radiated. Of that there is nothing but a bare memory. Santiago has spread out like a demented octopus, extending its eager tentacles in every direction; today five and a half million people live there, surviving however they can. It would be a pretty city, because it's well cared for, clean, and filled with gardens, if it didn't sit under a dark sombrero of pollution that in wintertime kills infants in their cradles, old people in nursing homes, and birds in the air. Santiaguinos have become accustomed to following the daily smog index just as faithfully as they keep track of the stock market or the soccer results. On days when the index climbs too high, the volume of vehicles allowed to circulate is restricted according to the number on the license plate, children don't play sports at school, and the rest of the population tries to breathe as little as possible. The first rain of the year washes the grime from the atmosphere and falls like acid over the city. If you walk outside without an umbrella you will feel as if lemon juice has been squirted in your eyes, but don't worry, no one has been blinded yet. Not all days are like that, sometimes the day dawns with a clear sky and you can appreciate the magnificent spectacle of snow-capped mountains.

There are cities, like Caracas or Mexico City, where poor and rich mix, but in Santiago the lines of demarcation are clear. The distance between the mansions of the wealthy on the foothills of the cordillera, with guards at the gate and four-car garages, and the shacks of the proletarian population where fifteen people live crowded together in two rooms without a bath, is astronomical. Every time I go to

Santiago I notice that part of the city is in black-and-white and the other in Technicolor. In the city center and in the worker's districts everything seems gray; the few trees that survive are exhausted, the walls faded, the clothing of the inhabitants very worn, even the dogs that wander among the garbage cans are mutts of indefinite color. In middle-class neighborhoods there are leafy trees, and the houses are modest but well cared for. In the areas where the wealthy live only the vegetation can be appreciated: the mansions are hidden behind impenetrable walls, no one walks down the streets, and the dogs are mastiffs let out only at night to guard the property.

Summer in the capital is long and hot. A fine, yellowish dust blankets the city during those months; the sun melts the asphalt and affects the mood of the inhabitants, so anyone who can tries to get away. When I was a girl, my family went for two months to the beach, a true safari in my grandfather's automobile, loaded with a ton of bundles on the luggage rack and three totally carsick children inside. At that time the roads were terrible and we had to snake up and down hills, which strained the vehicle to the breaking point. We always had to change tires once or twice, a task that entailed unloading all the bundles. My grandfather carried a huge pistol in his lap, like the ones used when people still fought duels, because he thought that bandits lurked on the Curacaví Hill, appropriately called the Graveyard. If there were highwaymen, they were probably just drifters who would have cut and run at the sound of the first shot, but just in case, we prayed as we drove past the hill—

undoubtedly an infallible protection against assault, since we never saw the famous *bandidos.* Nothing of that nature exists today. Now you can drive to seaside resorts in less than two hours, with excellent highways all the way. Until recently the only bad roads were those that led to the areas where the wealthy summer, part of their fight to preserve their exclusive beaches. They are horrified when they see the hoi polloi arriving in buses on the weekends with their dark-skinned children, their watermelon and roast chicken, and their radios and boom boxes blaring popular music— which is why they kept the dirt roads in the worst possible state. That has changed. As a rightist senator pontificated, "When democracy gets democratic, it doesn't work at all." The country is connected by one long artery, the Pan American and Austral Highways, and by an extensive network of paved and very safe roads. No guerrillas on the lookout for someone to kidnap, or gangs of drug traffickers defending their territory, or corrupt police looking for bribes, as in other Latin American countries rather more interesting than ours. You are much more likely to be mugged in the heart of the city than on a little-traveled road in the country.

Almost as soon as you leave Santiago, the countryside becomes bucolic: poplar-lined pastures, hills, and vineyards. To the visitor I recommend stopping to buy fruit and vegetables in the stands along the highway, or to take a little detour and drive into the villages and look for the house where you see a white cloth fluttering; there they serve leavened bread, honey, and eggs the color of gold.

Along the coastal route there are beaches, picturesque little villages, and coves where fishermen anchor their boats and spread their nets. There you find the fabulous treasures of our cuisine: first of all, the conger, king of the sea, wearing its jacket of jeweled scales; then the corbina, with its succulent white meat, accompanied by a court of a hundred other more modest but equally savory fish. Then comes the chorus of our shellfish: spider crabs, oysters small and large, mussels, abalone, langoustine, sea urchins, and many others, including some with such a questionable appearance that no foreigner dares try them, like the *pícoroco,* iodine and salt, pure marine essence. Our fish are so delicious that to prepare them you don't even need to know how to cook. You arrange a bed of minced onion in the bottom of a clay platter or Pyrex baking dish, lay the fresh fish, dotted with butter and sprinkled with salt and pepper and swimming in lemon juice, over the onion. Bake the fish in a hot oven until done—but not too long, you don't want it to get dry. Serve with one of our chilled white wines in the company of your closest friends.

Every year in December we would go with my grandfather to buy the Christmas turkey, which the campesinos raised for that holiday. I can see that old man, hobbling along on his bad leg, chasing around a field trying to catch the bird in question. He had to time his leap perfectly to fall on it, press it to the ground, and hold it while one of us struggled to bind its feet with a cord. Then he had to give the campesino a tip to kill the turkey out of sight of us children, otherwise we would have refused to taste it once it was cooked. It's very difficult to cut the throat of some creature with which you've established a personal relationship, as we could attest from the time my grandfather brought home a young goat to fatten in the patio of our house and roast on his birthday. That goat died of old age. And as it turned out, it wasn't a nanny but a male, and as soon as it grew horns, it attacked us at will.

The Santiago of my childhood had the pretensions of a large city but the soul of a village. Everything was public knowledge. Did someone miss mass on Sunday? That news traveled fast, and by Wednesday the parish priest was knocking at the door of the sinner to find out the reason. Men were stiff with hair pomade, starch, and vanity; women wore hat pins and kid gloves; elegant dress was expected when going into "the city" or to a movie—which people still thought of as a "talkie." Few houses had a refrigerator—in that my grandfather's house was very modern—and every day a hunchbacked man came by to deliver blocks of ice in sawdust for the neighborhood iceboxes. Our refrigerator, which ran for forty years without a repair, was fitted with a motor as noisy as a submarine, and from time to time shook the house with fits of coughing. The cook had to use a broom to fork out the bodies of electrocuted cats that had crawled beneath it to get warm. In the long run, that was a good method of birth control because dozens of cats were born on the roof tiles, and if some hadn't been zapped by the refrigerator we would have been inundated.

In our house, as in every Chilean home, there were animals. Dogs are acquired in different ways: inherited, received as a gift, picked up after they've been run over but not killed, or because they followed a child home from school, after which there's not a chance of throwing them out. This has always been the case and I hope it never changes. I don't know a single normal Chilean who ever bought a dog; the only people who do that are the fanatics from the Kennel Club, but no one takes them seriously. Almost all the dogs in Chile are called Blackie, whatever their color, and cats bear the generic names of Puss or Kitty; our family pets, however, always had Biblical names: Barrabas, Salome, Cain, except for one dog of dubious lineage whom we called Chickenpox because he appeared during an epidemic of that disease. Gangs of ownerless dogs roam the cities and towns of my country, not in the form of the hungry, miserable packs you see in other parts of the world but, rather, as organized communities. They are mild-mannered animals, satisfied with their social lot, a little lackadaisical. Once I read a study in which the author maintained that if all existing breeds of dogs were liberally intermingled, within a few generations they would narrow down to one type: a strong, astute beast of medium size, with short, wiry hair, a pointed muzzle, and willful tail: that is, the typical Chilean stray. I suppose we will come to that, and I hope also that with time we will succeed in fusing all human races; the result will be a rather short individual of indefinite color, adaptable, resilient, and resigned to the ups and downs of existence, like us Chileans.

In those days we went twice a day to the corner bakery to buy bread, and brought it home wrapped in a white cloth. The aroma of that bread just out of the oven, still warm, is one of the most tenacious memories of my childhood. Milk was a foamy cream sold from a tin can. A little bell that hung from the neck of the horse, and the smell of the stable invading the street, announced the arrival of the milk cart. Maids lined up with their bowls and basins and bought what was needed by the cup, which the milkman measured out by thrusting his hairy arm up to the armpit into large tin cans that were always swarming with flies. Sometimes several liters extra were bought to make *manjar blanco,* also called *dulce de leche,* a kind of blancmange that lasted several months when stored in the cool shadows of the cellar, where the home-bottled wine was also kept. First a fire of kindling and charcoal was built in the patio. A tripod was set over it that supported an iron kettle black from use. The ingredients were added in proportions of four cups of milk to one of sugar, and that mixture was flavored with two vanilla beans and the peel of a lemon and then boiled patiently for hours, occasionally stirred with a long wooden spoon. We children would watch from a distance, waiting for the process to end and the sweet to cool so we could lick the kettle. We were not allowed to come anywhere near it during the cooking; every time we would be told the sad story of the greedy little boy who fell into the pot and, as the tale went, "was dissolved in the boiling milk till not even his bones could be found." When pasteurized milk in bottles was invented, housewives dressed in their best clothes to be photographed—Hollywood-style—beside the white truck that replaced the unsanitary cart. Today not only are there whole, skim, and flavored milks, you can also buy bottled *manjar blanco,* no one makes it at home anymore.

In the summertime, little kids used to come by the house with baskets of blackberries and bags of quince for making preserves. The muscleman Gervasio Lonquimay also came by to check the metal springs of the cots and wash the wool mattresses, a task that could last three or four days because the wool had to dry in the sun and then be combed by hand before being stuffed back into the ticking. It was rumored that Gervasio Lonquimay had been in jail for slitting a rival's throat, gossip that lent him an aura of unquestionable prestige. The maids always offered him a cool drink for his thirst and towels for his sweat.

An organ grinder, always the same one, was a fixture in the streets of our barrio until one of my uncles bought his hurdy-gurdy and pathetic parrot, and went around cranking out music as the bird distributed little papers that brought good luck, to the horror of my grandfather and the rest of the family. I understand that my uncle's intention was to seduce a cousin with this display, but that the plan did not achieve the desired result: the girl married in a whirlwind and ran as far away as possible. Finally my uncle gave away the instrument but kept the parrot. It was very ill-humored, and at the first sign of inattention would nip a piece from the finger of anyone who came too close, but my uncle liked it because it swore like a corsair. It lived with him for thirty years, and who knows how many it had lived before: a Methuselah with feathers. Gypsy women, too, passed through the barrio, bamboozling the unwary with their mangled Spanish and those irresistible eyes that had seen so much of the world; they came always in twos or threes, with a half dozen runny-nosed brats clinging to their skirts. We were terrified of them because people said they stole little children, locked them in cages so they would grow up deformed, and then sold them to the circuses as freaks. They cast the evil eye on anyone who didn't give them money. They were thought to have magical powers: they could make jewels disappear without touching them, and unleash plagues of lice, warts, baldness, and rotted teeth. Even so, we couldn't resist the temptation to have them read the future in our palms. They always told me the same thing: a dark, mustached man would take me far away. Since I don't remember a single lover who fit that description, I have to assume they were referring to my stepfather, who had a mustache like a walrus and took me to many countries in his journeys as a diplomat.

AN OLD ENCHANTED HOUSE

My first memory of Chile is of a house I never knew, the protagonist of my first novel, *The House of the*

Spirits, where it appears as the large home that shelters the issue of the Truebas. That fictional family bears an alarming resemblance to my mother's; I could never have invented such a clan. Actually, I had no need to, with a family like mine you don't need imagination. The idea of the "large old house on the corner," so much a part of the novel, evolved from the home on Calle Cueto where my mother was born, and so frequently evoked by my grandfather that it seems I have lived there. There are no houses like that left in Santiago, they've been devoured by progress and demographic growth, but they still exist in the provinces. I can see it: vast and drowsy, worn by use and abuse, with high ceilings and narrow windows and three patios, the first with orange trees and jasmine and a singing fountain, the second with a weed-choked garden, and the third a clutter of laundry utensils, dog houses, chicken coops, and unhealthful quarters for the maids, like cells in a dungeon. To go to the bathroom at night, you had to make an excursion with a flashlight, defying cold air and spiders and turning a deaf ear to the sounds of creaking wood and scurrying mice. That huge old house, which had an entrance on two streets, was one-story tall with a mansard roof, and it harbored a tribe of great-grandparents, maiden aunts, cousins, servants, poor relatives, and guests who became permanent residents; no one tried to throw them out because in Chile "visitors" are protected by the sacred code of hospitality. There was also an occasional ghost of dubious authenticity, always in plentiful supply in my family. Some attest that souls in pain wandered within those walls, but one of my older relatives confessed to me that as a boy he dressed up in an ancient military uniform to frighten Tía Cupertina. That poor maiden lady hadn't the slightest doubt that her nocturnal visitor was the spirit of Don José Miguel Carrera, one of the fathers of the nation, who had come to ask for money to say masses for the salvation of his warrior's soul.

My maternal aunts and uncles, the Barros, were twelve rather eccentric brothers and sisters, though none was hopelessly mad. When they married, some stayed on in that house on Calle Cueto with their spouses and children. That is what my grandmother Isabel did when she married my grandfather Agustín. The couple not only lived in that chicken yard of outlandish relatives but, on the death of my great-grandparents, they bought the house and for several years raised their four children there. My grandfather modernized the house, but his wife suffered from asthma because of the damp; in addition, the poor moved into the neighborhood and "the best people" began to emigrate en masse to the eastern part of the city. Bowing to social pressure, my grandfather built a modern house in the barrio of Providencia, and although it was then on the city's outskirts, he predicted that the area would prosper. The man had a good eye, because within a few years Providencia had become the most elegant residential area in the capital, though that ended long ago when the middle class began to creep up the slopes of the hills and the truly rich moved farther and farther up the cordillera, where the condors nest. Today Providencia is a chaos of traffic, commerce, offices, and restaurants, where only the old live in ancient apartment buildings, but then it was bordered by open country where wealthy families had their summer farms, and where the air was clear and life bucolic. I will have more to say about that house a little later, but for the moment, let's go back to my family.

Chile is a modern country of fifteen million inhabitants, but the residue of a tribal mentality lingers on. Things haven't changed much, despite the demographic explosion, especially in the provinces, where each family stays within its tight circle, large or small. We are divided into clans that share an interest or an ideology, and their members resemble one another, dress similarly, think and act like clones, and, of course, protect one another, excluding anyone not of the group. I can mention, for example, clans of agricultural landholders (I'm referring to the owners, not humble campesinos), doctors, politicians (regardless of party), entrepreneurs, soldiers, teamsters, and then all the rest. Above even the clan is the family, inviolable and sacred; no one escapes his duties to family. For example, my stepfather Tío Ramón calls from time to time to tell me that some uncle three times removed, whom I have never met, has died and left a daughter in a difficult situation. The girl wants to study nursing but doesn't have the means to do so. It is up to Tío Ramón, as the elder of the clan, to contact anyone who has blood ties to the deceased, from close relatives to far-flung cousins, to finance the education of the future nurse. To refuse to help would be so despicable that it would be writ in the annals of the family for several generations. Given the importance that family has for us, I have chosen mine as the thread that ties this book together, so if

I expand on a member of my clan it's with good reason—though at times that may be nothing more than my wish not to lose those blood ties that bind me, too, to my land. My relatives will serve to illustrate certain vices and virtues of the Chilean character. As a scientific method this may be questionable, but from the literary point of view it has its advantages.

My grandfather, who came from a small family ruined by the early death of his father, fell in love with a girl famed for her beauty, Rosa Barros, but the girl died mysteriously before they could wed. All that remains of her is a pair of sepia-tone photographs, faded in the fog of time, in which her features are barely perceptible. Years later, my grandfather married Isabel, Rosa's younger sister. In those days in Santiago, everyone within a specific social class knew each other, so that marriages, though not arranged as they were in India, were indeed family matters. Therefore it seemed logical to my grandfather that since he had been accepted by the Barros as a suitor of one of their daughters, there was no reason why he should not court the other.

My grandfather Agustín was a slim man when he was young; he had a distinctive aquiline nose, and, solemn and proud, wore a black suit cut from one of his dead father's. He came from an old family of Spanish-Basque origins, but unlike his relatives, he was poor. His family didn't offer a great deal to talk about, except for Tío Jorge, my uncle, who was as elegant and good-looking as a prince; he had a brilliant future before him, and was desired by all the señoritas of marrying age. He had the bad fortune, however, to fall in love with a woman *de medio pelo,* as Chileans call the struggling lower middle class. In another country they might have been able to love one another without tragedy, but in the world they lived in they were condemned to being ostracized. This woman adored my Tío Jorge for fifty years, but she wore a moth-eaten fox stole, she dyed her hair carrot red, she smoked up a storm, and she drank beer from the bottle, more than enough reasons for my great-grandmother Ester to declare war on her and forbid her son to speak his beloved's name in her presence. He obeyed without a word, but the day after the matriarch's death, he married his lover, who by then was a mature woman with lung problems, although still captivating. They loved each other in their poverty and no one was ever able to part them. Two days after he died of a heart attack, they found her dead in bed, wrapped in her husband's old bathrobe.

Allow me a word about that great-grandmother Ester, because I believe that her powerful influence explains some aspects of the character of her descendants, and in no little measure she represents the intransigent matriarch who was, and is, so common in the culture. The mother figure reaches mythological proportions in our country, so I don't find my Tío Jorge's submission surprising. Jewish and Italian mothers are dilettantes compared with the Chilean ones. I have just discovered, by chance, that her husband had a bad head for business and lost his lands and the fortune he had inherited; it seems that his creditors were his own brothers. When he realized he was ruined, he went out to his country house and blew a hole in his chest with a shotgun. I say I just learned about this because for a hundred years the family hid that story, and it is still mentioned in whispers. Suicide was considered a particularly opprobrious sin, since the body couldn't be buried in the consecrated earth of a Catholic cemetery. To escape that shame, my great-grandfather's relatives dressed his corpse in a morning coat and top hat, sat him in a horse-drawn carriage, and drove him to Santiago, where he could be given a Christian burial because everyone, including the priest, turned a blind eye. This event divided the direct descendants, who swear that the story of the suicide is calumny, and the descendants of the dead man's brothers, who ended up with his wealth. In either case, the widow was left sunken in depression and poverty. She had been a happy, pretty woman, a piano virtuoso, but upon her husband's death she dressed in severe mourning, locked the piano, and from that day forward left her home only to go to daily mass. Over time, arthritis and obesity turned her into a monstrous statue trapped within the four walls of her house. Once a week the parish priest served communion at the house. That somber widow instilled in her children the idea that the world is a vale of tears and that we come here only to suffer. A prisoner in her wheelchair, she made judgments on the lives of others; nothing escaped her tiny falcon eyes and her prophet's tongue. For the filming of *The House of the Spirits,* to play that role they had to transport an actress the size of a whale from England to the studio in Copenhagen, after removing several seats in the plane to contain her unimaginable corpulence. She appears on the screen for only an instant, but she makes a memorable impression.

Unlike Doña Ester and her descendants, so solemn and serious, my maternal relatives were happy-go-lucky, exuberant, spendthrift womanizers, quick to bet on the horses, play music, and dance the polka. Now, dancing is not typical behavior among Chileans, who as a rule lack any sense of rhythm. One of the great discoveries I made in Venezuela, where I went to live in 1975, is the therapeutic power of dance. You get three Venezuelans together and one will play the drums or the guitar and the other two will dance; there is no ill that can resist that treatment. Our parties, in contrast, seem like funerals: the men gather in a corner to talk business and the women die of boredom. Only the young dance, seduced by North American music, but as soon as they marry they turn solemn like their parents. The greater part of the anecdotes and characters in my books are based on that unique family. The women were delicate, spiritual, and amusing. The men were tall, handsome, and always game for a fistfight. They were also *chineros,* which is what they called habitués of brothels, and more than one died of "an undiagnosed illness." I must assume that the culture of the whorehouse is important in Chile because it appears again and again in the literature, as if our authors were obsessed with it. Even though I don't consider myself an expert on the subject, I am not innocent of creating a whore with a heart of gold; mine, from my first novel, is named Tránsito Soto.

I have a hundred-year-old aunt who aspires to sainthood, and whose only wish has been to go into the convent, but no congregation, not even the Little Sisters of Charity, could tolerate her for more than a few weeks, so the family has had to look after her. Believe me, there is nothing as insufferable as a saint, I wouldn't sic one on my worst enemy. During the Sunday lunches at my grandfather's house, my uncles laid plans to murder her, but she always escaped unharmed, and is alive to this day. In her youth, this woman wore a habit of her own invention, sang religious hymns for hours in her angelic voice, and, the minute no one was watching, slipped out to Calle Maipú to shout at the top of her lungs for the salvation of the ladies of the night, who welcomed her with a rain of rotten vegetables. On that same street, my Tío Jaime, my mother's cousin, earned money for his medical studies by pumping an accordion in "houses of ill repute." At dawn, at the top of *his* lungs, he would be singing a song titled "I Want a Naked Woman," creating such an uproar that the good sisters would come outside to protest. In those days, the black list of the Catholic Church included books like *The Count of Monte Cristo,* so imagine the furor that wanting a naked woman would cause when yodeled by my uncle. Jaime became the most famous and most beloved pediatrician in the country, and the most picturesque politician—quite likely to recite his speeches in the Senate in rhymed verse— and by far the most radical of my relatives, a Communist to the left of Mao, when Mao was still in diapers. Today he is a handsome and lucid old man who wears fiery-red socks as a symbol of his political beliefs. Another of my relatives used to take off his trousers in the street to give them to the poor, and a photograph of him in his undershorts, though properly attired in hat, jacket, and tie, appeared more than once in newspapers. He had such an exalted idea of himself that in his will he left instructions that he wanted to be buried standing up: that way when he knocked at the gates of heaven he could look God directly in the eye.

I was born in Lima, where my father was one of the secretaries at the embassy. The reason I grew up in my grandfather's house in Santiago is that my parents' marriage was a disaster from the beginning. One day when I was four, my father went out to buy cigarettes and never came back. The truth is that he didn't start out to buy cigarettes, as everyone always said, but instead went off on a wild spree disguised as a Peruvian Indian woman and wearing bright petticoats and a wig with long braids. He left my mother in Lima with a pile of unpaid bills and three children, the youngest a newborn baby. I suppose that that early abandonment made some dent in my psyche, because there are so many abandoned children in my books that I could found an orphanage. The fathers of my characters are dead, have disappeared, or are so distant and authoritarian they might as well live on another planet. When she found herself without a husband and on her own in a strange country, my mother had to conquer the monumental pride with which she'd been brought up and go home to live with my grandfather. My first years in Lima are obliterated in the mists of lost memory; all my recollections of childhood are linked to Chile.

I grew up in a patriarchal family in which my grandfather was like God: infallible, omniscient, and omnipotent. His house in the barrio of Providencia wasn't a shadow of the house my great-grandparents had on Calle Cueto, but for me, during my early years, it was my universe. Not long ago, a Japanese newspaperman went to Santiago, intending to photograph the supposed "large house on the corner" that appears in my first novel. It was pointless to try to explain to him that it was fictional. At the end of such a long journey, the poor man suffered a terrible disappointment. Santiago has been demolished and rebuilt several times. Nothing lasts in this city. The home my grandfather built is today an unprepossessing discothèque, a depressing mélange of black plastic and psychedelic lights. The residence on Calle Cueto was demolished many years ago and in its place stand modern towers for low-income housing, unrecognizable among so many dozens of similar buildings.

Please allow me a commentary about that demolition, as a sentimental whim. One day the machines of progress arrived with the mission of pulverizing the large home of my ancestors, and for weeks the implacable iron dinosaurs tore into structures with their great claws. When finally the Sahara-like dust storms subsided, the passersby saw, to their amazement, that a few palm trees had survived intact. Solitary, denuded, with their scruffy manes and air of ashy beggars, they awaited their end. Instead of the feared executioner, however, sweaty workmen carrying picks and shovels appeared and, working like an army of ants, dug trenches around each tree, loosening them from the earth. Those slender trees held handfuls of soil in their threadlike roots. Cranes bore the wounded giants to deep holes gardeners had prepared in a different spot, and planted them there. The trunks moaned quietly, the leaves drooped in yellow strands, and for a while it seemed that nothing could save them from their agony, but they were tenacious. A slow subterranean rebellion fought to preserve life, vegetal tentacles spread out, blending clumps of dirt from Calle Cueto with new soil. With the inevitable arrival of spring, the palms awoke, swaying from the waist, shaking their hair, rejuvenated despite their trauma. The image of those trees from the home of my ancestors often comes to mind when I think of my destiny as an expatriate. It is my fate to wander from place to place, and to adapt to new soils. I believe I will be able to do that because handfuls of Chilean soil are caught in my roots; I carry them with me always. In any case, the Japanese newspaperman who traveled to the end of the world to photograph a house from a novel returned home with empty hands.

My grandfather's house was like my uncle's, and like the house of any family of similar circumstances. Chileans are not noted for originality: inside, their houses are all more or less the same. I'm told that now the wealthy contract decorators and buy the hardware for their bathrooms abroad, but in those days no one had ever heard of interior decoration. In the living room, which was swept by inexplicable drafts, there were heavy, plush, oxblood-colored drapes, teardrop chandeliers, an out-of-tune grand piano, and a large grandfather clock, black as a coffin, which struck the hours with funereal sonority. There were also two horrific French porcelain figurines of damsels with powdered wigs and gentlemen in high heels. My uncles used them to tune their reflexes: they threw them headfirst at each other in the vain hope they would be dropped and would shatter into a thousand pieces. This dwelling was inhabited by eccentric humans, half-wild pets, and my grandmother's ghostly friends, who had followed her from the house on Calle Cueto and who, even after she died, continued to wander through the rooms.

My grandfather Agustín was as solid and strong as a warrior, even though he was born with one leg shorter than the other. It never occurred to him to consult a doctor about this problem, he preferred to go to a "bonesetter." This was a blind man who treated the legs of injured horses at the Riding Club, and who knew more about bones than any orthopedic surgeon. Over time, my grandfather's lameness worsened, it caused arthritis, and threw his vertebral column out of line, so that every movement was torture, but I never heard him complain about his pains or his problems— though like any respectable Chilean he complained about everything else. He bore the torment of his poor skeleton with the help of aspirin and long draughts of water. Later I learned that this wasn't innocent water but gin, which he drank neat, like a pirate, with no effect on his behavior or his health. He lived nearly a century with never a sign of a single loose screw. His pain did not excuse him from his duties as a gentleman, and to the end of his days, when he was nothing but a bundle of old bones and leather, he laboriously got up from his chair to greet and bid farewell to a woman.

On my desk I have a photograph of my grandfather. He looks like a Basque peasant. He's in profile, wearing a black beret that accentuates his aquiline nose and the firm expression of a face marked by deep furrows. He grew old strengthened by intelligence and reinforced by experience. He died with a full head of white hair and blue eyes as piercing as those of his youth. "How hard it is to die," he told me one day when he was already very weary of pain. He spoke in proverbs, he knew hundreds of folk tales, and recited long poems from memory. This formidable man gave me the gift of discipline and love for language; without them I could not devote myself to writing today. He also taught me to observe nature and to love the landscape of Chile. He always said that just as Romans live among ruins and fountains without seeing them, we Chileans live in the most dazzling country on the planet without appreciating it. We don't notice the quiet presence of the snowy mountains, the sleeping volcanoes, and the unending hills that wrap us in their monumental embrace; we are not amazed by the frothing fury of the Pacific bursting upon our coasts, or the quiet lakes of the south and their musical waterfalls; we don't, like pilgrims, venerate the millenary nature of our native-growth forests, the moonscape of the deserts of the north, the fecund Araucan rivers, or the blue glaciers where time is shattered into splinters.

We're talking about 1950. How long I've lived, my God! Getting old is a drawn-out and sneaky process. Every so often, I forget that time is passing because inside I'm still not thirty, but inevitably my grandchildren confront me with the harsh truth when they ask me if "in your day" we had electricity. These same grandchildren insist that there's a country inside my head where the characters in my books live their lives. When I tell them stories about Chile, they think I'm referring to that invented place.

A MILLEFEUILLE PASTRY

W_{ho} are we, we Chileans? It's difficult for me to define us in writing, but from fifty yards I can pick out a compatriot with one glance. I find them everywhere. In a sacred temple in Nepal, in the Amazon jungle, at Mardi Gras in New Orleans, on the brilliant ice of Iceland, there you will find some Chilean with his unmistakable way of walking or her singing accent. Although because of the length of our narrow country we are separated by thousands of kilometers, we are tenaciously alike; we talk the same tongue and share similar customs. The only exceptions are the upper class, which has descended with little distraction from Europeans, and the Indians—the Aymaras and a few Quechuas in the north and the Mapuches in the south—who fight to maintain their identities in a world where there is constantly less space for them.

I grew up with the story that there are no problems of race in Chile. I can't understand how we dare repeat such a falsehood. We don't talk in terms of "racism" but, rather, of "the class system" (we love euphemisms), but there is little difference between them. Not only do racism and/or class consciousness exist, they are as deeply rooted as molars. Whoever maintains that racism is a thing of the past is dead wrong, as I found out in my latest visit, when I learned that one of the most brilliant graduates in the law school was denied a place in a prestigious law firm because "he didn't fit the corporate profile." In other words, he was a mestizo, that is, he had mixed blood, and a Mapuche surname. The firm's clients would never be confident of his ability to represent them; nor would they allow him to go out with one of their daughters. Just as in the rest of Latin America, the upper class of Chile is relatively white, and the farther one descends the steep social ladder the more Indian the characteristics become. Nevertheless, lacking other points of comparison, most of us consider ourselves white. It was a surprise for me to discover that in the United States I am a "person of color." (Once, when I was filling out a form, I opened my blouse to show my skin color to an Afro-American INS officer who was intent on placing me in the last racial category on his list: "Other." He didn't seem to think it was funny.)

Although few pure Indians remain—approximately ten

percent of the population—their blood runs through the veins of our mestizo people. Mapuches are rather short, and generally have short legs, a long torso, dark skin, dark hair and eyes, and prominent cheekbones. They have an atavistic—and justifiable—mistrust for all non-Indians, whom they call *huincas,* which doesn't translate as "whites" but as "land robbers." These Indians, who are divided into several tribes, contributed greatly to forming the national character, although there was a time when no one with any self-respect admitted the least association with them because of their reputation for drinking, laziness, and thieving. That was not, however, the opinion of Don Alonso de Ercilla y Zúñiga, a renowned Spanish soldier and writer who came to Chile in the middle of the sixteenth century. Ercilla is the author of *La Araucana,* an epic poem about the Spanish conquest and the fierce resistance of the natives. In the prologue, he addresses the king, his lord, saying of the Araucans that

> With undiluted courage and stubborn determination they reclaimed and sustained their freedom, spilling in sacrifice for it so much blood, both theirs and that of Spaniards, that in truth it may be said that there are few places left unstained with it, or not covered with bones. . . . And the loss of life is so great, because of all those who have died in this endeavor, that to fill out their numbers and to swell their ranks, their women also come to war, and, fighting at times like males, they deliver themselves spiritedly unto death.

Some Mapuche tribes have rebelled in recent years, and the country cannot ignore them much longer. In fact, they are somewhat in vogue. Intellectuals and ecologists scramble to come up with some lance-toting ancestor to adorn their genealogical tree; a heroic Indian in a family tree is much more glamorous than some decadent marquis softened by life at court. I confess that I myself have tried to adopt a Mapuche name, so I could puff myself up about having a chieftain great-grandfather—the same way that in the past people bought titles of European nobility—but to date I've been unsuccessful. I suspect that that's how my father obtained his coat of arms: three starving mutts on a blue field, as I recall. The escutcheon in question was hidden away in the cellar and never mentioned, because titles of nobility were abolished when Chile declared its independence from Spain. In Chile there is nothing so ridiculous as to try to pass as a nobleman. When I worked at the United Nations I had a boss who was a true Italian count, and he had to change his calling cards because of the guffaws his heraldry provoked.

Indian chieftains won their rank with superhuman feats of strength and valor. They would hoist a tree trunk from those virgin forests to their shoulders, and whoever could support that weight longest became the *toqui*. As if that weren't enough, they recited an extemporaneous composition without pausing for breath, because in addition to proving their physical capabilities they had to make the case for the coherence and beauty of their words. That may be the source of our age-old vice of poetry. From that moment, no one would dispute his authority until the next tournament. No torture contrived by ingenious Spanish conquistadors, however horrible, succeeded in demoralizing those formidable dark-skinned heroes, who died without a moan impaled on pikes, quartered by four horses, or slowly burned alive over hot coals. Our Indians did not have a significant culture like those of the Aztec, Maya, or Inca empires; they were rough, primitive, bad tempered, and small in number, but so brave that they were in a state of war for three hundred years, first against the Spanish colonizers, then against the republic. They were pacified in 1880, and not much was heard from them for more than a century, but now the Mapuches—"people of the earth"—have again taken up the fight because what little land is still theirs is threatened by construction of a dam on the Bío Bío River.

The artistic and cultural products of our Indians are as somber as everything else produced in the country. They color their cloth and weavings with vegetable dyes: dark red, black, gray, white; their musical instruments are as lugubrious as the song of whales; their dances are dull, monotonous, and last so long that in the end they bring rain. Their craftwork is beautiful, but it lacks the exuberance and variety of the art of Mexico, Peru, or Guatemala.

The Aymaras—"children of the sun"—are very different from the Mapuches; they are the same Indians as those in Bolivia, and they come and go, ignoring boundaries, because that region has been theirs forever. They are affable by nature and although they have maintained their customs, their language, and their beliefs, they have been integrated into the culture of the whites, especially in matters relating to commerce. In that they are different from some groups of Quechuas in the most isolated areas of the mountains of Peru, for whom the government is the enemy, just as it was during the Spanish colonial period. The war of independence and the creation of the Republic of Peru have not modified their lives in any way; in fact some are not even aware that those things happened.

The unfortunate Indians of Tierra del Fuego, in the extreme southern tip of Chile, died by gunfire and disease long ago; of those tribes only a handful of Alacalufes survive. Hunters were paid a bounty for every pair of ears they brought as proof of having killed an Indian; in this way the territory was successfully "cleared" for the colonists. These Indians were giants who lived, nearly naked, in an inclement, icy zone where only seals can take comfort.

African blood was never incorporated into Chilean stock, which would have given us rhythm and beauty; neither was there, as there was in Argentina, significant Italian immigration, which would have made us extroverted, vain, and happy; there weren't even enough Asians, as there were in Peru, to compensate for our solemnity and spice up our cuisine. But I am sure that if enthusiastic adventurers had converged from the four corners of the earth to people our land, proud Spanish–Basque families would have managed to intermingle with them as little as possible, unless they were northern Europeans. It has to be said: our immigration policy has been openly racist. For a long time we didn't accept Asians, blacks, or anyone with a deep tan. It occurred to one president in the eighteen hundreds to bring Germans from the Black Forest and grant them land in the south, which of course wasn't his to give, since it belonged to the Mapuches, but no one noted that detail except the legitimate owners. The idea was that Teutonic blood would set a fine example for our mestizos, instilling in them a work ethic, discipline, punctuality, and organization. The brown skin and coarse hair of the Indians were looked down on; a few German genes wouldn't hurt us a bit, was the thinking of the authorities of the time. It was hoped that these admirable immigrants would marry Chilean women and from that mixture white blood would win out over that of the humble Indians—which is what happened in Valdivia and Osorno, provinces that today can boast of tall men, full-breasted women, blue-eyed children, and authentic apple strudel. Color prejudice is so strong that if a woman has yellow hair, even if she has the face of an iguana, men turn to look at her in the street. My own hair was colored from the time I was a tiny child, using a sweet-smelling liquid called Bay Rum; there is no other explanation for the miracle that the lank, black strands I was born with were transformed before I was six months old into angelic golden curls. Such extremes weren't necessary with my brothers, because one had curly hair and the other was blond. In any case, the people who emigrated from the Black Forest have been very influential in Chile, and according to the opinion of many, they rescued the south from barbarism and made it the splendid paradise it is today.

After the Second World War, a different wave of Germans

came to seek refuge in Chile, where there was so much sympathy for them that our government didn't affiliate with the Allies until the last moment, when it was impossible to remain neutral. During the war the Chilean Nazi Party paraded with brown uniforms, flags bearing swastikas, and arms raised in a Nazi salute. My grandmother ran alongside, throwing tomatoes at them. This woman was an exception because in Chile people were so anti-Semitic that the word *Jew* was a dirty word, and I have friends who had their mouths washed out with soap for having dared say it. When you referred to that people, you said *Israelite* or *Hebrew,* nearly always in a whisper. There still exists today a mysterious colony out in the countryside called Dignidad, a Nazi camp that is completely out of bounds, as if it were an independent nation; no government has been able to dismantle it because it is believed that it has the covert protection of the armed forces. During the days of the dictatorship (1973–1989), Dignidad was a torture center used by the intelligence services. Today its chief is a fugitive from justice, accused of raping minors, among other crimes. The people in the area welcome these supposed Nazis, however, because they staff an excellent hospital, which they place at the service of the local population. At the entrance to the colony there is a stupendous German restaurant where they have the best pastries for miles around, served by strange blond men with facial tics, who speak in monosyllables and have lizard eyes. This I haven't witnessed, but have been told.

During the nineteenth century, the English arrived in large numbers and took control of maritime and rail transport, along with the import and export industry. Some third- or fourth-generation descendants, who had never set foot on English soil but nevertheless called it home, took pride in speaking Spanish with an accent and in keeping up with the news by reading out-of-date papers from the "homeland." My grandfather, who had many business dealings with companies that raised sheep in Patagonia for the British textile industry, always said that he never signed a contract, that a man's word and a handshake were more than enough. The English—*gringos,* as we generically call anyone who has blond hair or whose mother tongue is English—established schools and clubs and taught us various extremely boring games, including bridge.

We Chileans like the Germans for their sausage, their beer, and their Prussian helmets, as well as the goose step our military adopted for parades, but in practice we try to emulate the English. We admire them so much that we think we're the English of Latin America, just as we believe that the English are the Chileans of Europe. During the ridiculous war in the Falkland Islands (1982), instead of backing the Argentines, who are our neighbors, we supported the British, and from that time forward the then Prime Minister of Great Britain, Margaret Thatcher, has been the soul mate of General Pinochet. Latin America will never forgive us for such a faux pas. It's clear that we have much in common with the fair-haired sons of Albion: individualism, good manners, a sense of fair play, class consciousness, bad teeth, and austerity. (British austerity does not, of course, include the royalty, who are to the English what Las Vegas is to the

Mojave Desert). We are fascinated by the eccentricity of which the British tend to boast, but are incapable of imitating it, because we are too afraid of ridicule; on the other hand, we do try to copy their apparent self-control. I say apparent because in certain circumstances—for example, a soccer match—English and Chileans lose their heads equally and are capable of drawing and quartering their opponents. In the same fashion, despite their reputation for being level-headed, both can exhibit fierce cruelty. The atrocities committed by the English throughout their history are matched by those Chileans commit as soon as they have a good excuse—or impunity. Our history is spattered with examples of barbarism. It isn't for nothing that our motto is "By Right or by Force," a phrase that has always seemed particularly stupid to me. In the nine months of the revolution of 1891, more Chileans died than during the four years of the war against Peru and Bolivia (1879–1883), many of them shot in the back or tortured, others thrown into the sea with stones tied to their ankles. The technique of "disappearing" ideological enemies, which several Latin American dictatorships were so strongly committed to during the seventies and eighties, had been practiced in Chile nearly a century earlier. None of which takes away from the fact that our democracy was the most solid, and the oldest, on the continent. We were proud of the efficacy of our institutions; of our incorruptible *carabineros,* our police; of the uprightness of our judges; and of the fact that no president became rich while in power: just the opposite, they often left the Palacio de la Moneda—ironically, the Palace of Money—poorer than they came in. Following 1973 we have not had occasion to boast about those things.

In addition to the English, Germans, Arabs, Jews, Spaniards, and Italians, immigrants from Central Europe made their way to our shores: scientists, inventors, academics, some true geniuses, all of whom we refer to, without distinction as "Yugoslavians."

After the Spanish Civil War, refugees came to Chile escaping the defeat. In 1939, the poet Pablo Neruda, at the direction of the Chilean government, chartered a ship, the *Winnipeg,* which sailed from Marseilles carrying a cargo of intellectuals, writers, artists, physicians, engineers, and fine craftsmen. The affluent families of Santiago came to Valparaíso to meet the boat and offer hospitality to the voyagers. My grandfather was one of them; there was always a place set at his table for Spanish friends who showed up unannounced. I hadn't been born yet, but I grew up hearing stories of the Civil War and the salty songs those passionate anarchists and republicans used to sing. Those people shook the country from its colonial doldrums with their ideas, their arts and professions, their suffering and passions, their extravagant ways. One of those refugees, a Catalan friend of my family, took me one day to see a linotype. He was a thin, nervous young man with the profile of an angry bird; he never ate vegetables because he considered them burro fodder, and he lived obsessed with the idea of returning to Spain when Franco died, never suspecting that the man would live another forty years. This friend was a typographer by trade and smelled of a mixture of garlic and ink. From the far end of the table, I used to watch him pick at his food and rail against Franco, monarchists, and priests; he never glanced in my direction because he detested children and dogs equally. To my surprise, one winter day the Catalan announced that he was taking me for a walk. He threw his long muffler around his neck and we set off in silence. He took me to a gray building where we went through a metal door and walked down corridors stacked with enormous rolls of paper. A deafening noise shook the walls. I watched him being transformed: his step became lighter, his eyes gleamed, he smiled. For the first time, he touched me. Taking my hand, he led me to a fabulous machine, a kind of black locomotive with all its works in view, as if it had been violently gutted. He touched its keys and with a warlike roar the matrices fell into place, forming the lines of a text.

"Some damned German clockmaker who emigrated to the United States patented this marvel in 1884," he yelled into my ear. "It's called a linotype, *line of types*. Before that, you had to compose the text by setting the type by hand, letter by letter."

"Why 'damned'?" I asked, yelling back.

"Because twelve years earlier, my father invented the same machine and set it up in his patio, but no one gave a fig," he replied.

The typographer never returned to Spain, he stayed and operated the word machine, married, children fell from the skies, he learned to eat vegetables, and he adopted several generations of stray dogs. He gave me the memory of the linotype and a taste for the smell of ink and paper.

In the society I was born into, in the forties, there were unbreachable barriers between the social classes. Today those lines are more subtle, but they're there, as eternal as the Great Wall of China. Climbing the social ladder was once impossible; descending was more common—sometimes the only nudge needed was to move or to marry badly, which did not mean to a cad or heartless person but someone beneath you. Money had little to do with it. Just as you didn't slip to a lower class when you lost your money, neither did you rise a notch by amassing a fortune, a lesson learned by many rich Arabs and Jews who were never accepted in the exclusive circles of "decent people." This was how those who found themselves at the top of the social pyramid referred to themselves (assuming, naturally, that all the rest were "*in*decent people").

Foreigners rarely catch on to how this shocking class system operates because social interchange is polite and friendly at every level. The worst epithet bestowed on the military who took over the government in the seventies was that they were "boosted-up *rotos*." My aunts said that there was nothing tackier than being a Pinochet adherent. They said that not as a criticism of his dictatorship, with which they were in full accord, but in regard to class status. Now, few people dare use the word *roto* in public, because it's considered bad form, but most have it on the tip of their tongues. Our society is like a millefeuille pastry, a thousand layers, each person in his place, each in her class, every person marked by birth. People introduced themselves—and this is still true in the upper class—using both surnames, in order to establish their identity and lineage. We Chileans have a well-trained eye for determining a person's place in society by physical appearance, color of skin, mannerisms, and especially the way of speaking. In other countries, accents vary from place to place; in Chile they change according to social class. Usually we can also immediately determine the subclass, of which there are at least thirty, determined by different levels of tastelessness, social ambition, vulgarity, new money, and so on. You can tell, for example, where a person belongs by the resort he goes to in the summer.

The process of automatic classification we Chileans practice when we are introduced has a name, *situating,* and is the equivalent of what dogs do when they sniff each other's hindquarters. Since 1973, the year of the military coup that changed so many things, situating has become a little more complex because in the first three minutes of conversation you also have to guess whether the person you're speaking to was for or against the dictatorship. Today very few confess they were in favor, but even so it's a good idea to establish a political orientation before you express extreme opinions. The same is true among Chileans who live outside the country, where the obligatory question is, When did you leave? If he, or she, says before 1973, it means that person is a rightist and was fleeing Allende's socialism; if he left between 1973 and 1978, you can be sure he is a political refugee; but any time after that, and she may be an "economic exile," which is how those who left Chile looking for job opportunities are qualified. It is more difficult to place those who stayed in Chile, partly because those individuals learned to keep their opinions to themselves.

SIRENS SCANNING THE SEA

No one asks a returning Chilean where he's been or what he saw; on the other hand, we immediately inform the foreigner arriving for a visit that our women are the most beautiful in the world, that our flag won some mysterious international contest, and that our climate is idyllic. Judge for yourself: the flag is nearly identical to that of Texas, and the most notable aspect of our climate is that while there's a drought in the north there are sure to be floods in the south. And when I say floods, I am talking Biblical deluges that leave hundreds dead, thousands injured, and the economy in ruins; they do, however, trigger that solidarity that tends to bog down in normal times. We Chileans are enchanted by states of emergency. In Santiago the temperatures are worse than in Madrid; in summer we die of the heat and in winter of the cold, but no one has air conditioning or decent heating, because that would be tantamount to admitting that the climate isn't as good as they say it is. When the air gets too agreeable, it's a sure sign that there's going to be an earthquake. We have more than six hundred volcanoes, some where the petrified lava of former eruptions is still hot, others with poetic Mapuche names: Pirepillán, demon of the snow; Petrohué, land of the mists. From time to time these sleeping giants rouse themselves from their dreams with a long bellow, and then it seems as if the end of the world has come. Experts on earthquakes say that sooner or later Chile will disappear, buried in lava or dragged to the bottom of the sea by one of those gigantic waves that tend to rise up in fury in the Pacific, but I hope this doesn't discourage potential tourists, because the probability that it will happen precisely during their visit is rather remote.

The matter of female beauty requires a separate comment. It's outrageous flattery raised to a national level. The truth is that I have never heard it said outside the country that Chilean women are quite as spectacular as my amiable compatriots assert. Our women are no more alluring than Venezuela's, who win all the international beauty contests, or Brazil's, who sashay along the beaches parading their *café au lait* curves, to mention only two of our rivals. But according to popular Chilean mythology, from time immemorial sailors have deserted their ships, entranced by the longhaired sirens who wait, scanning the sea, on our beaches. This monumental approbation on the part of our men is so gratifying that we women are inclined to forgive them many things. How can we deny them when they find us beautiful? If there is a thread of truth in all this, perhaps it is that a Chilean woman's attraction lies in a blend of strength and flirtatiousness that few men can resist—that's according to what I hear, for it hasn't been a hundred percent true in my case. My male friends tell me that the amorous game of glances, of suggestion, of giving a man his head and then reining him in, is what captivates them, but I suppose that wasn't invented in Chile, we imported it from Andalusia.

For several years I worked for a women's magazine where we were constantly surrounded with the most sought-after models and the latest candidates for the Miss Chile competition. The models, in general, were so anorexic that most of the time they sat perfectly motionless, staring straight ahead, like turtles, which made them very attractive since any man passing by could imagine that they were stupefied by the sight of him. In any case, these beauties all looked like tourists. Without exception, the blood flowing through their veins was European: they were tall, slim, and had light hair and eyes. That is not the typical Chilean woman, the one you see in public: a mestizo, brunette and rather short—although I can't deny that recent generations are taller. Today's young people seem gigantic to me (admittedly, I am barely five feet tall . . .). Nearly all the female characters in my novels are inspired by Chilean women whom I know very well because I worked with them and for them for several years. More than by upper-class señoritas, with their long legs and blond manes, I've been impressed by the women of the people: mature, strong, hard-working, earthy. In their youth they are passionate lovers, and afterward they are the pillars of their family, good mothers and good companions to men who often do not deserve them. Under their wings they harbor their own and others' children, friends, relatives, and hangers-on. They are always bone tired, weary from serving others, always putting off what they should do for themselves; the last among the last, they work tirelessly and age prematurely, but they never lose their capacity to laugh at themselves, their romantic hope that their partners will change, or the small flame of rebelliousness that burns in their hearts. Most are martyrs by vocation: they are the first up to wait on their families and the last to go to bed; they take pride in suffering and sacrificing. They sigh and weep with great gusto as they tell one another the stories of abuse from husband and children!

Chilean women dress simply, nearly always in slacks; they wear their hair down and use little makeup. On the beach or at a party they all look the same, a chorus of clones. I took the time to go through old magazines, from the end of the sixties to today, and I find that in this sense very little has changed in forty years. I think that the only difference is the volume among various hairstyles. Every woman has "a little black dress," which is synonymous with elegance and which, with few variations, accompanies her from puberty to coffin. One of the reasons I don't live in Chile is that I wouldn't fit in. My closet has enough veils, plumes, and glitter to outfit the entire cast of *Swan Lake;* furthermore, I have tinted my hair every color chemicals have to offer, and have never stepped out of the bathroom without my eye makeup. Being permanently on a diet is a symbol of status among us, though in more than one poll the men interviewed have used terms like "soft, curvy, with something you can get a grip on," to describe how they prefer their women. We don't believe them: surely they say that to console us . . . which is why we cover our protuberances with long sweaters or starched blouses, just the opposite from Caribbean women, who proudly display their pectoral abundance in low necklines and posteriors sheathed in fluorescent spandex. But beauty is a matter of attitude. I remember one woman with a Cyrano de Bergerac nose. In view of her lack of success in Santiago, she went to Paris and in no time at all she had appeared in France's most sophisticated fashion magazine—eight pages, full color—wearing a turban and in bold profile! From that time, this woman-attached-to-a-nose has passed into posterity as a symbol of the crowed-over beauty of Chilean women.

Some frivolous thinkers believe that Chile is a matriarchy, deceived perhaps by the strong personality of its women, who seem to carry the lead in society. They are free and well organized, they keep their maiden names when they marry, they compete head to head in the workforce and not only manage their families but frequently support them. They are more interesting than most men, but that does not affect the reality: they live in an unyielding patriarchy. To begin with, a woman's work or intellect isn't respected; we must work twice as hard as any man to earn half the recognition. Don't even mention the field of literature! But we're not going to talk about that, because it's bad for my blood pressure. Men have the economic and political power, which is passed from one male to the next, like the baton in a relay, while women, with few exceptions, are pushed to the side. Chile is a macho country: there is so much testosterone floating in the air that it's a miracle women don't grow beards.

There is no secret about machismo in Mexico; it's in their *rancheras,* their country ballads, but among us it is much more veiled—though no less injurious. Sociologists have traced the causes back to the Spanish conquest, but since male dominance is a world problem, its roots must be much more ancient, it isn't fair to blame only the Spaniards. At any rate, I will repeat what I've read about it. The Araucan Indians were polygamous and treated women very badly; they would abandon them, and their children, and leave as a group to look for new hunting grounds, where they took new women and had more children, whom they left in turn. The mothers took care of their offspring as best they could, a custom that in a way persists in the psyche of our people. Chilean women tend to accept—though not forgive—abandonment by their men because they think of it as an endemic ill, something inherent in the male nature. As for the Spanish conquistadors, very few of them brought women with them, so they coupled with Indian women, whom they valued far less than a horse. From these unequal unions were born humiliated daughters who would themselves be raped as women, and sons who feared and admired the soldier father: bad-tempered, unjust, master of all rights, including those of life and death. As those sons grew up, they identified with their fathers, never with the conquered race of the mother. Some conquistadors had as many as thirty concubines, not counting the women they raped and immediately abandoned. The Inquisition railed against the Mapuches for their polygamous customs, but overlooked the harems of captive Indian women accompanying the Spaniards: more mestizo children meant more subjects for the crown of Spain and more souls for the Christian religion. From those violent embraces come our peoples, and to this day men act as if they were on horseback surveying the world from on high, giving orders, conquering. As a theory, that isn't half bad, right?

Chilean women are abettors of machismo: they bring up their daughters to serve and their sons to be served. While on the one hand they fight for their rights and work tirelessly, on the other, they wait on their husband and male children, assisted by their daughters, who from an early age are well instructed regarding their obligations. Modern girls are rebelling, of course, but the minute they fall in love they repeat the learned pattern, confusing love with service. It makes me sad to see splendid girls waiting on their boyfriends as if they were invalids. They not only serve the meal, they offer to cut the meat. It makes me unhappy because I was the same way. Not long ago a TV comic, a man dressed as a woman, scored a great hit by imitating a model wife. Poor Elvira—that was his name—ironed shirts, cooked complicated meals, did the children's homework, waxed the floor by hand, and flew around to put on nice clothes and makeup before her husband came home from work, so he wouldn't find her ugly. Elvira never rested, and everything was always her fault. One time she even ran a marathon behind the bus her husband was taking to work, to hand him the briefcase *he* had forgotten. The program made men howl with laughter, but it bothered the women so much that finally it was taken off the air; wives didn't like seeing themselves portrayed so faithfully by the ineffable Elvira.

My American husband, who takes responsibility for half the chores in our house, is scandalized by Chilean machismo. When a man washes the plate he's eaten from, he considers that he's "helping" his wife or mother, and expects to be praised for his effort. Among our Chilean friends there is always some woman who'll serve breakfast in bed to adolescent boys, wash their clothes, and make their bed. If there's no *nana,* the mother or a sister does it, something that would seldom happen in the United States. Willie was also horrified by the institution of the maid. I prefer not to tell him that in the past the duties of these women were even more intimate, although that was never discussed openly: mothers looked the other way and the fathers boasted of their sons' backstairs feats. He's a tiger, they would say, remembering their own experiences, a "chip off the old block." The general idea was for the boy to satisfy his sexual needs with the maid, so he wouldn't "go too far" with a girl of his own social class; and after all, a maid was safer than a prostitute. In rural areas there was a local version of the Spanish *derecho a pernada,* which in feudal times allowed the lord to bed any bride on the night of her wedding. In Chile, the tradition was never that organized: the *patron* just went to bed with anyone and at any time he pleased. So the landowners sowed their lands with bastards, and even today there are regions where nearly everyone has the same last name. (One of my ancestors knelt to pray after every rape: "Lord, I don't do this for fun and games, only for more sons to serve in Your name . . .") Today the *nanas* have become so emancipated that the lords of their domains prefer to hire illegal immigrants from Peru, whom they can mistreat as badly as they used to their Chilean servant girls.

In matters of education and health, Chilean women are at or above the level of the men, but not in opportunities and political power. The normal pattern in the workforce is that they do the hard work and the men direct. Very few women occupy high posts in government, industry, or private or public enterprise; they bump into that ever-present glass ceiling. When a woman does reach a top-level position, let's say, minister in the government or director of a bank, it is cause for amazement and admiration. In the last ten years, however, public opinion is registering positive for women as political leaders: they are seen as a viable alternative because they have demonstrated that they are often more honest, efficient, and hardworking than men. What a revelation! When women organize, they wield great influence, but they seem unaware of their own strength. There was the example, for instance, during the administration of Salvador Allende, when rightist women went out beating pots and pans to protest shortages and to dump chicken feathers in front of the Military School, inciting the soldiers to subversion. They helped foment the military coup. Years later, women were the first to go out and publicly denounce military repression, confronting water hoses, nightsticks, and bullets. They formed a powerful group called "Women for Life," which played a fundamental role in overturning the dictatorship of Pinochet, but after the election they decided to dissolve the movement. Once more they ceded their power to men.

I should clarify that Chilean women, who are so slow to fight for political power, are true guerrillas when it comes to love. In affairs of the heart, they are truly dangerous, and, it must be said, they fall in love with considerable frequency. According to the statistics, 58 percent of married women are unfaithful. I wonder if couples don't often switch: while the man seduces his friend's wife, his own spouse is in the same hotel in the arms of his friend. In colonial times, when Chile was part of the vice-regency of Lima, the Inquisition sent a Dominican priest from Peru to accuse a number of women of high social standing of engaging in oral sex with their husbands. (And how did they know *that?*) The trial never went anywhere because the women in question refused to be browbeaten. The night after the trial they sent their husbands—who somehow or other must have participated in the sin, though only the women were being judged—to dissuade the inquisitor. They overtook him in a dark, narrow street and without further ado they castrated him, like a steer. The poor Dominican returned to Lima *sans* testicles, and the matter was never mentioned again.

Though not reaching quite such extremes, I have a story about a friend who couldn't rid himself of an impassioned lover until finally one day he escaped while she was taking a siesta. He had packed a few belongings in an overnight bag and was running down the street after a taxi when he felt something like a bear crash onto his shoulders, throwing him to the ground, where he lay squashed like a cockroach. It was his lover, who had charged after him, completely naked and screeching like a banshee. People ran from houses all over the neighborhood to enjoy the spectacle. The men were amused, but as soon as the women realized what was going on, they helped the "wronged" woman hold down my slippery friend, and then among them they lifted him up and carted him back to the bed he'd abandoned during siesta time.

I could give three hundred additional examples, but surely these are enough.

PRAYING TO GOD

The account I just gave you about those ladies of the colonial era, the ones who defied the Inquisition, marks an exceptional moment in our history because in reality the power of the Catholic Church is irrefutable, and now with the strength of fundamentalist Catholic movements like the Opus Dei and the Legionarios de Cristo behind it, that power is even more unassailable.

Chileans are very religious, although in practice that

has a lot more to do with fetishism and superstition than with mystic restiveness or theological enlightenment. No Chilean calls him or herself an atheist, not even dyed-in-the-wool communists, because the term is considered an insult. The word *agnostic* is preferred, and usually even the strongest nonbelievers are converted on their deathbeds since they risk too much if they don't, and a last-hour confession never hurt anyone. This spiritual compulsion rises from the earth itself: a people who live amid mountains logically turn their eyes toward the heavens. Manifestations of faith are impressive. Convoked by the Church, thousands and thousands of young people carrying candles and flowers march in long processions giving praise to the Virgin Mary or praying for peace at a deafening decibel level, screaming with the enthusiasm teenagers in other countries exhibit at rock concerts. It used to be enormously popular to say the rosary as a family, and the celebrations of the month of Mary always scored a great success, but recently the television soaps have boasted more fans.

As you might expect, an esoteric strain runs through my family. One of my uncles has spent seventy years of his life preaching about the encounter with nothingness. He has many followers. If I had paid attention to him when I was young, I wouldn't be studying Buddhism today and trying fruitlessly to stand on my head in my yoga class. When it comes to matters of holiness, however, that poor demented hundred-year-old woman who disguised herself as a nun and tried to reform the prostitutes on Calle Maipú can't hold a candle to my great-aunt who sprouted wings. They weren't wings with golden feathers, like those of Renaissance angels, that would have attracted everyone's attention; they were discreet little stumps on her shoulders, erroneously diagnosed by doctors as a bone deformity. Sometimes, depending on where the light was coming from, we could see a halo like a plate of light floating above her head. I recounted her drama in the *Stories of Eva Luna,* and I don't want to repeat it here. It's enough to say that in contrast with the Chilean's general tendency to complain about everything, she was always content, even though she had a tragic fate. In another person that attitude of unfounded happiness would have been unpardonable, but in such a transparent woman it was easy to tolerate. I have always kept her photograph on my desk so I will recognize her when she slips slyly into the pages of a book or appears in some corner of the house.

In Chile there is a plethora of saints of all stripes, which isn't strange, considering it is the most Catholic country in the world—more Catholic than Ireland, and certainly much more so than the Vatican. A few years ago we had a young girl, in appearance very like the statue of Saint Sebastian the Martyr, who performed amazing cures. The press and television swarmed all over her, as well as multitudes of pilgrims who never gave her a moment's peace. When she was examined more closely, she turned out to be a transvestite, but that did nothing to detract from her prestige or put an end to her marvels. Just the opposite. Every so often we wake up with the announcement that another saint or new Messiah has made his or her appearance, which never fails to attract hopeful throngs. In the seventies, when I was working as a journalist, I happened to write an article about a girl who was credited with having the gift of prophecy and a faculty for curing animals and restoring dead engines to working order. The humble little house where she lived was filled with the country folk who came every day, always at the same hour, to witness her discreet miracles. They swore that they heard an inexplicable rain of rocks on the roof of the hut, rattling like the end of the world, and that the earth would tremble and the girl would fall into a trance. I had the opportunity to attend a couple of these events, and I witnessed the trance, during which the young saint displayed the extraordinary physical strength of a gladiator, but I don't recall any rocks from the skies or quaking earth. It's possible, as a local evangelical preacher explained, that these things failed to occur because I was there, a skeptic capable of ruining even the most legitimate miracle. No matter, the phenomenon was reported in the newspapers, and people's interest in the saint kept rising until the army came and put an end to everything in its own way. I used her story ten years later in one of my novels.

Catholics form a majority in Chile, although there are more and more Evangelicals and Pentecostals who irritate everyone because they have a direct understanding with God while everyone else must pass through the priestly bureaucracy. The Mormons, who are also numerous and very powerful, serve their followers as a valuable employment agency, the way that members of the Radical Party used to do. Whoever is left is either Jewish, Muslim, or, in my generation, a New Age spirtualist, which is a cocktail of ecological, Christian, and Buddhist practices, along with a few rituals recently rescued from the Indian reservations, and with the usual accompaniment of gurus, astrologists, psychics, and other spiritual guides. Since the health care system was privatized and pharmaceuticals became an immoral business, folkloric and Eastern medicine, *machis* or *meicas,* our Indian healers, and self-taught herbalists and purveyors of miraculous cures have in part replaced traditional medicine, with equally effective results. Half of my friends are in the hands of a psychic who controls their destinies and keeps them safe by cleansing their auras, laying on hands, or leading them on astral journeys. The last time I was in Chile, I was hypnotized by a friend who is studying to be a *curandero,* a healer, who led me back through several incarnations. It wasn't easy to return to the present, however, since my friend hadn't reached that part of the course, but the experiment was well worth the effort because I discovered that in former lives I was *not* Genghis Khan, as my mother believes.

I have not succeeded in completely shaking free of religion, and when I'm faced with any difficulty, the first thing that occurs to me is to pray, just in case, which is what all Chileans do, even atheists . . . forgive me, agnostics. Let's say I need a taxi. Experience has taught me that once through the Lord's Prayer will make one appear. There was a time, somewhere between infancy and the age of fifteen, when I nursed the fantasy of being a nun as a way of disguising the fact that I most surely was not going to find a husband, and to this day I haven't completely discarded that fancy. I am still assailed by the temptation to end my days in poverty, silence, and solitude in a Benedictine order or a Buddhist convent. Theological subtleties are not what count with me, what I like is the lifestyle. Despite my unconquerable frivolity, the monastic life attracts me. When I was fifteen, I left the church forever and acquired a horror of religions in general and monotheistic faiths in particular. I am not alone in this predicament; many women my age, guerrillas in the battle for women's lib, are similarly uncomfortable in patriarchal religions—can you think of one that isn't?—and they have had to invent their own cults, although in Chile even cults have a Christian bent. However animist someone may claim to be, there will always be a cross somewhere in her house, or around her neck. My religion, should anyone be interested, can be reduced to a simple question: What is the most generous thing one can do in this case? If that question doesn't apply, I have another: What would my grandfather think about this? None of which relieves me of the compulsion to cross myself in my hour of need.

I used to say that Chile is a fundamentalist country, but after seeing the excesses of the Taliban, I have to moderate my opinion. Maybe we're not fundamentalists, but we're close. We have been fortunate in that in Chile, unlike other countries in Latin America, the Catholic Church—with a few regrettable exceptions—has almost always been on the side of the poor, which has gained it enormous respect and sympathy. During the dictatorship, many priests and nuns took on the task of helping the victims of repression, and they paid dearly for it. As Pinochet said in 1979, "the only persons going around crying for democracy to be restored in Chile are the politicians and one or two priests." That was the period when the generals posited that Chile was blessed with "a totalitarian democracy."

Churches are filled on Sundays, and the pope is venerated, although no one pays any attention to his views on contraceptives because it's thought that there's no way an aged celibate who doesn't have to work for a living can be an expert on that subject. Religion is colorful and ritualistic. We don't have Carnival, but we do have processions. Every saint is noted for his or her special power, like the gods on Olympus: restoring sight to the blind, punishing unfaithful husbands, finding a sweetheart, protecting drivers. The most popular, however, is undoubtedly Padre Hurtado, who isn't a saint as yet, though we all hope he soon will be, no matter that the Vatican is not noted for swift action. This amazing priest founded a center for good works called "The Home of Christ," which today is a multimillion-dollar enterprise devoted entirely to aiding the poor. Padre Hurtado is so miraculous that I have seldom asked him for something that hasn't been granted, after I have made some significant sacrifice or contributed a fair sum to his charitable works. I must be one of the few people alive who have read the three complete volumes of the ageless epic *La Araucana,* in rhyme and old Spanish. I didn't do it out of curiosity, or to pretend to be cultured, but to fulfill a promise to Padre Hurtado. This man of good heart maintained that a moral crisis is produced when the same affluent Catholics who faithfully go to mass deny their workers a dignified wage. These words should be engraved on the thousand-peso note, so we never forget them.

There are also various representations of the Virgin Mary, which compete among themselves: those faithful to the Virgen del Carmen, patron saint of the armed forces, believe that the Virgen de Lourdes or the Virgen de La Tirana are inferior, a sentiment returned with equal delicacy by the devout followers of the latter Virgins. Regarding La Tirana, it's of interest to mention that in the summertime her festival is celebrated in a sanctuary near the city of Iquique, in the north of Chile, where various followers dance in her honor. The fiesta is a little like Brazil's Carnival, but on a much more sedate scale: as I've said before, we Chileans are not the extroverts of Latin America. The dance studios prepare all year for the festival, practicing choreographed dances and making costumes, and on the scheduled day dancers perform before the statue of La Tirana, the men made up as heroes, like Batman, for example, and the girls wearing revealing blouses, skirts that barely cover the buttocks, and boots with high heels. It is not too surprising, therefore, that the Church does not sanction these demonstrations of popular faith.

Not satisfied with a huge and multihued bevy of saints, we also have a delicious oral tradition of evil spirits, interventions of the devil, and dead who rise from their tombs. My grandfather swore that he saw the devil on a bus, and that he recognized him because he had green cloven hooves like a billygoat.

. . .

In Chiloé, a group of islands off Puerto Montt in the south of the country, they tell tales of warlocks and malicious monsters: of La Pincoya, a beautiful damsel who rises from the water to trap unwary men, and the *Caleuche,* an enchanted ship that carries away the dead. On nights of the full moon, glowing lights indicate sites where treasures are hidden. It is said that in Chiloé there was for a long time a government of warlocks called the *Recta Provincia,* or Righteous Province, which met in caves by night. The guardians of those caves were the *inbunches,* fearsome creatures that feed on blood, and whose bones have been broken, and eyelids and anuses stitched shut, by witches. The Chilean's imagination for cruelty never ceases to terrify me.

Chiloé's culture is different from that of the rest of the country, and their people are so proud of their isolation that they oppose the construction of a bridge that will join the large island to Puerto Montt. It is such an extraordinary place that every Chilean and every tourist must visit it at least once, even at the risk of staying forever. The Chilotes live as they did a hundred years ago, dedicated to agriculture and the fishing industry, specifically salmon. Buildings are constructed solely of wood, and in the heart of each house there is always a huge wood stove burning day and night for cooking and for providing warmth to the family, friends, and enemies gathered around it. The scent of those houses in winter is an ineradicable memory: blazing, aromatic firewood, wet wool, soup kettles. The Chilotes were the last to cast their lot with the republic when Chile declared its independence from Spain, and in 1826 they tried to join with the crown of England. They say that the *Recta Provincia* attributed to warlocks was in fact a shadow government in times when the inhabitants refused to accept the authority of the Chilean republic.

My grandmother Isabel didn't believe in witches, but I wouldn't be in the least surprised to learn that she had attempted to fly on a broomstick because she spent her life practicing effects with paranormal phenomena and trying to communicate with the Great Beyond, an activity that in her time the Catholic Church regarded with a jaundiced eye. Somehow that good lady managed to attract mysterious forces that moved the table during her séances. Today that table is in my home, after having traveled around the world several times, following my stepfather in his diplomatic career only to be lost during the years of exile. My mother recovered it through some burst of inspiration and shipped it to me in California, by air freight. It would have been cheaper to send an elephant because we are talking here about a massive, carved-wood Spanish table that has a formidable foot at the center formed of four ferocious lions. It takes three men to lift it. I don't know what trick my grandmother performed when she made it dance around the room by stroking it lightly with her index finger. That lady convinced her descendants that after her death she would come to visit whenever they summoned her, and I suppose she has kept her promise. I don't claim that her ghost, or any other, is at my side every day—I expect that they have more important matters to attend to—but I like the idea that she is ready to come in case of some compelling need.

That good woman maintained that we all have psychic powers but since we don't use them they atrophy, like muscles, and finally disappear. I must clarify that her paranormal experiments were never a macabre experience, none of the dark rooms, mortuary candelabra, and organ music that we connect with Transylvania. Telepathy, the ability to move objects without touching them, seeing the future, and communicating with souls in the Great Beyond may happen any hour of the day, and in a very casual manner. For example, my grandmother didn't believe in telephones, which in Chile were a disaster until the day of the cell phone, and used telepathy instead to send recipes for apple pie to the three Morla sisters, her bosom friends in the Hermandad Blanca, Sisterhood in White, who lived on the other side of the city. Whether or not the method worked was left unproved, because all four were terrible cooks. The Hermandad Blanca was composed of those eccentric ladies and my grandfather, who was a total nonbeliever but nevertheless insisted on accompanying his wife so he could protect her in case of danger. The man was a skeptic by nature, and never was persuaded that the souls of the dead moved the table, but once his wife suggested that it might not be spirits but extraterrestrials, he embraced the idea enthusiastically because he considered that a more scientific explanation.

There is nothing strange in all this. Half of Chile is guided by the horoscope, by seers, or by the vague prognostications of the I Ching; the other half hang crystals around their necks or follow feng shui. On the lovelorn advice programs on TV, problems are resolved with tarot cards. The greater part of former militant leftist revolutionaries are now dedicated to spiritual practices. (There is some dialectic link between the guerrilla mentality and the esoteric that I can't quite put my finger on.) My grandmother's sessions seem more rational to me than vows made to saints, buying indulgences to guarantee heaven, or pilgrimages to local holy women: buses bursting with people and stands selling sausages and miraculous color prints. I have often heard that my grandmother moved the sugar bowl without touching it, using only her mental powers. I'm not sure whether I witnessed that feat or if from hearing it so often I've convinced myself it's true. I don't remember the sugar bowl, but it seems to me there was a little silver bell, topped with an effeminate prince, that was used in the dining room to call the servants between courses. I don't know if I dreamed the episode, if I invented it, or if it truly happened: I see that little bell slide silently across the tablecloth, as if the prince had taken on a life of his own, make a stunning Olympian turn, to the amazement of the diners, and return to my grandmother's place at the foot of the table. This happens with many events and anecdotes in my life: it seems I have lived them, but when I write them down in the clear light of logic, they seem unlikely. That really doesn't disturb me, however. What does it matter if these events happened or if I imagined them? Life is, after all, a dream.

. . .

I did not inherit my grandmother's psychic powers, but she opened my mind to the mysteries of the world. I accept that anything is possible. She maintained that there are multiple dimensions to reality, and that it isn't prudent to trust solely in reason and in our limited senses in trying to understand life; other tools of perception exist, such as instinct, imagination, dreams, emotions, and intuition. She introduced me to magical realism long before the so-called boom in Latin American literature made it fashionable. Her views have helped me in my work because I confront each book with the same criterion she used to conduct her sessions: calling on the spirits with delicacy, so they will tell me their lives. Literary characters, like my grandmother's apparitions, are fragile beings, easily frightened; they must be treated with care so they will feel comfortable in my pages.

Apparitions, tables that move on their own, miraculous saints and devils with green hooves riding on public transport make life and death more interesting. Souls in pain know no borders. I have a friend in Chile who wakes up at night to find tall, skinny visitors from Africa dressed in tunics and armed with spears, specters only he can see. His wife, who sleeps right beside him, has never seen the Africans, only two eighteenth-century English gentlewomen who walk through doors. And another friend of mine lived in a house in Santiago where lamps mysteriously crashed to the floor and chairs overturned; the source of the mayhem was discovered to be the ghost of a Danish geographer who was dug up in the patio along with his maps and his notebook.

How did that poor wandering soul end up so far from home? We will never know, but the fact is that after several novenas and a few masses for him, the geographer left. He must have been a Calvinist or a Lutheran during his lifetime and didn't like the papist rites.

My grandmother claimed that space is filled with presences, the dead and the living all mixed together. It's a fabulous idea, and that's why my husband and I have built a large house in northern California with high ceilings, beams, and arches that invite ghosts from various periods and latitudes, especially those of the far south. In an attempt to replicate my great-grandparents' large house, we have aged it through the costly and laborious process of attacking the doors with hammers, staining the walls with paint, rusting the iron with acid, and treading on the plants in the garden. The result is rather convincing: I believe that more than one distraught spirit might settle in with us, deceived by the look of the property. During the process of adding centuries to the house, the neighbors watched from the street, open-mouthed, not understanding why we were building a new house if we wanted an old one. The reason is that in California you don't find much in the way of Chile's colonial style, and in any case, nothing is truly old. Don't forget that before 1849 there was no San Francisco. Where it stands now was a village called Yerba Buena; it was populated by a handful of Mexicans and Mormons, and its only visitors were fur dealers. It was gold fever that brought the hordes to San Francisco. A house that looks like ours is a historical impossibility in these parts.

THE LANDSCAPE OF
CHILDHOOD

It is very difficult to determine what a typical Chilean family is like, but I can say, without any fear of contradiction, that mine was *not* average. Nor was I a typical señorita with regard to the mores of the milieu in which I grew up; I made a clean getaway, as they say. I will describe some parts of my youth, to see whether in the process I shed some light on aspects of my country's society, which in those days was much less tolerant than it is today—which says a lot about how it was then. The Second World War was a cataclysm that shook the world, and changed everything from geopolitics and science to customs, culture, and art. Without much discussion, new ideas swept away those the society had held for centuries, but innovations were slow to cross two oceans or to break through the impenetrable wall of the Andes. It took several years for new modes to reach Chile.

My clairvoyant grandmother died suddenly of leukemia. She didn't fight for life, she gave herself to death enthusiastically because she was very curious to see heaven. During her lifetime in this world she had the good fortune to be loved and protected by her husband, who bore her extravagant behavior with good humor; if he hadn't she would have ended up in a madhouse. I've read several letters she left in her own hand, in which she seems to be a melancholy woman with a morbid fascination with death. I remember her, however, as luminous and ironic, and full of gusto for life. Her leaving was like a catastrophic wind; the entire house went into mourning and I learned what it was to be afraid. I feared the devil that appeared in the mirrors, the ghosts that hovered in the corners, the mice in the cellar, I was terrified that my mother would die and I would have to go to an orphanage, that my father—that man whose name could not be spoken—would come back and take me away, I was afraid of committing sins and going to hell, afraid of the gypsies, of the bogeyman whose name the nursemaid invoked to threaten me . . . in short, I had an endless list, more than enough reasons to live in terror.

My grandfather, furious at being abandoned by the great love of his life, dressed in black from head to toe, painted the furniture the same color and forbade parties, music, flowers, and desserts. He spent the day at his office, lunched in town, dined at the Union Club, and on weekends played golf and jai alai, or went to the mountains to ski. He was one of the first to initiate that sport in days when getting to the runs was an odyssey equal to scaling Everest. He never imagined that one day Chile would be a Mecca of winter sports, where Olympic teams from all over the world are sent to train. We saw him only a minute in the early morning, but he was nonetheless a determining factor in my formation. Before we went to school, my brothers and I would go by to say good morning. He received us in his room filled with dark furniture and smelling of an English soap with the trademark Lifebuoy.

He never patted or hugged us—he thought it unhealthy—but we would go to any lengths for a word of approval from him. Later, when I was about seven and had begun to read the newspaper and ask questions, he noticed my presence, and then began a relationship that would continue long after his death, because till this day signs of his hand are perceptible in my character, and I am constantly nourished by the anecdotes he told me.

My childhood wasn't a happy one, but it was interesting. I was never bored, thanks to the books that belonged to my Tío Pablo, who at that point was still a bachelor living at home. He was an unreconstructed reader; his bookshelves covered the walls from floor to ceiling, and volumes piled up on the floor to be covered with dust and cobwebs. He stole books from his friends without a trace of guilt because he thought printed material—except what belonged to him—was the patrimony of all humankind. He let me read his treasures because he meant to pass his vice as a reader on to me, no matter what the cost. He gave me a doll when I finished reading *War and Peace,* a fat book with tiny print. There was no censorship in that house, but my grandfather did not allow lights to be on in my room after nine o'clock at night, and to circumvent that my Tío Pablo gave me a flashlight. My best memories of those years are of books I read beneath the covers, using my flashlight. We Chilean children read the novels of Emilio Salgari and Jules Verne, the *Treasury of Youth* and collections of didactic little novels that promoted obedience and purity as maximum virtues. We also read the magazine *El Peneca,* a reader that was published every Wednesday. As early as Tuesday I was stationed at the door to keep the magazine from falling into my brothers' hands first. I devoured that as an aperitif, then gobbled up more succulent dishes, such as *Anna Karenina* and *Les Misérables*. For dessert I savored fairy tales. Those magnificent books allowed me to escape the rather shabby reality of that house in mourning in which we children, like the cats, were considered a nuisance.

My mother was again an eligible young woman, thanks to having been able to have her marriage annulled, and though living under her father's wing, she had a few admirers— maybe one or two dozen by my calculations. Besides being beautiful, she had that ethereal and vulnerable look some girls had then, a look that's been completely lost in these days when ladies lift weights. Her fragility was very seductive because even the wimpiest man felt strong by her side. She was one of those women who make men want to protect her, exactly the opposite of me; I am more like a tank at full throttle. Instead of wearing black and weeping about being abandoned by her frivolous husband, as was expected of her, my mother tried to enjoy herself as much as she could under the circumstances, which was very little because in those days women couldn't go to a tearoom alone, to say nothing of the movies. Films that were in the least interesting were classified as "not recommended for señoritas," which meant that they could be seen only in the company of a man of the family, who took responsibility for the moral harm the spectacle could inflict upon a sensitive female psyche. A few snapshots have survived from those years; in them my mother looks like a younger sister of the actress Ava Gardner. She was born with beauty: luminous skin, easy laugh, classic features, and a great natural elegance, more than enough reason for sharp tongues to comment on her every move. If her platonic suitors stirred the sanctimonious society of Santiago, imagine the scandal that erupted when it learned of her love affair with a married man, the father of four children and nephew of a bishop.

Among her many suitors, my mother chose the ugliest. Ramón Huidobro resembled a green frog, but with the kiss of love he was transformed into a prince, just like the fairy tale, and now I can swear that he's handsome. Clandestine relationships had existed always, we Chileans are expert in that, but their romance had nothing clandestine about it, and soon was an open secret. Given the impossibility of either dissuading his daughter or preventing the scandal, my grandfather decided to defuse the gossip by bringing the lover to live beneath his roof, defying the church and all of society. The bishop called in person to set things straight, but my grandfather took his arm and in friendly fashion led him to the door, stating that he took care of his own sins and those of his daughter as well. With time, that lover would become my stepfather, the incomparable Tío Ramón, friend, confidant, my only and true father, but when he came to live in our house I thought he was my enemy, and I tried to make his life impossible. Fifty years later, he assures me that wasn't true, that I never declared war, but he says that out of a noble heart to salve my conscience, because I remember all too well my plans for his slow, painful death.

Chile is possibly the one country in the galaxy where there is no divorce, and that's because no one dares defy the priests, even though 71 percent of the population has been demanding it for a long time. No legislator, not even those who have been separated from their wives and partnered a series of other women in quick succession, is willing to stand up to the priests, and the result is that divorce law sleeps year after year in the "pending" file, and when finally it is approved it will be with so much red tape and so many conditions that it will be easier to murder your spouse than to divorce him or her. My best friend, tired of waiting for her marriage to be annulled, read the newspapers every day with the hope that she would see her husband's name. She never dared pray that the man would be dealt the death he deserved, but if she had asked Padre Hurtado sweetly, I have no doubt he would have complied. For more than a hundred years legal loopholes have allowed thousands of couples to annul their marriages. And that is what my parents did. All it took was my grandfather's determination and connections to have my father disappear by magic and my mother declared an unmarried woman with three illegitimate children, which our law calls "putative" offspring. My father signed the papers without a word, once he'd been assured that he wouldn't have to support his children. The process consists of having a series of witnesses present false testimony before a judge who pretends to believe what he's told. To obtain an annulment you must at least have a lawyer: not exactly cheap since he charges by the hour; his time is golden and he's in no hurry to shorten the negotiations. The necessary requirement, if the lawyer is to "iron out" the annulment, is that the couple must be in agreement because if one of the two refuses to participate in the farce, as my stepfather's first wife did, there's no deal. The result is that men and women pair and separate without papers of any kind, which is what nearly all the people I know have done. As I am writing these reflections, in the third millennium, the divorce law is still pending, even though the president of the republic annulled his first marriage and married a second time. At the rate we're going, my mother and Tío Ramón, who are already in their eighties and have lived together more than half a century, will die without being able to legalize their situation. It no longer matters to either of them, and even if they could marry they wouldn't; they prefer to be remembered as legendary lovers.

Like my father, Tío Ramón worked in the Ministry of Foreign Affairs, and shortly after being installed beneath my grandfather's protective roof in the role of illegal son-in-law, he was sent on a diplomatic mission to Bolivia. That was in the early fifties. My mother and all three of us children went with him.

Before I began to travel, I was convinced that all families were like mine, that Chile was the center of the universe, and that every human being looked like us and spoke Spanish as a first language: English and French were school assignments, like geometry. We had barely crossed the border when I had my first hint of the vastness of the world and realized that no one, absolutely no one, knew how special my family was. I quickly learned what it is to feel rejected. From the moment we left Chile and began to travel from country to country, I became the new girl in the neighborhood, the foreigner at school, the strange one who dressed differently and didn't even know how to talk like everyone else. I couldn't picture the time that I would return to familiar territory in Santiago, but when finally that happened, several years later, I didn't fit in there either, because I'd been away too long. Being a foreigner, as I have been almost forever, means that I have to make a much greater effort than the natives, which has kept me on my toes and forced me to become flexible and adapt to different surroundings. This condition has some advantages for someone who earns her living by observing; nothing seems natural to me, almost everything surprises me. I ask absurd questions, but sometimes I ask them of the right people and thus get ideas for my novels.

To be frank, one of the things that most attracts me to Willie is his challenging and confident attitude. He never has any doubt about himself or his circumstances. He has always lived in the same country, he knows how to order from a catalogue, vote by mail, open a bottle of aspirin, and where to call when the kitchen floods. I envy his certainty. He feels totally at home in his body, in his language, in his country, in his life. There's a certain freshness and innocence in people who have always lived in one place and can count on witnesses to their passage through the world. In contrast, those of us who have moved on many times develop tough skin out of necessity. Since we lack roots or corroboration of who we are, we must put our trust in memory to give continuity to our lives . . . but memory is always cloudy, we can't trust it. Things that happened in the past have fuzzy outlines, they're pale; it's as if my life has been nothing but a series of illusions, of fleeting images, of events I don't understand, or only half understand. I have absolutely no sense of certainty. Nor can I picture Chile as a geographic locale with certain precise characteristics: a real and definable place. I see it the way a country road might look as night falls, when the long shadows of the poplars trick our vision and the landscape is no more substantial than a dream.

A SOBER AND SERIOUS PEOPLE

A friend of mine says that we—we Chileans—may be poor, but that we have delicate feet. She's referring, of course, to our unjustified sensitivity, always just beneath the skin, to our solemn pride, to our tendency to become idiotically sober given the slightest opportunity. Where did such characteristics come from? I suppose they can be attributed, at least in part, to the mother country, Spain, which bequeathed us a mixture of passion and severity; another portion we owe to the blood of the long-suffering Araucans; and the rest we can blame on fate.

I have, through my father, a little French blood, and a touch of Indian—all you have to do is look at me to see that—but my heritage is primarily Spanish-Basque. The founders of families like mine tried to establish dynasties, and to do that they invented an aristocratic past, though in fact they were laborers and adventurers who came to the tail end of America with their hands out. Of blue blood, so to speak, not a drop. They were ambitious and hardworking, and they appropriated the most fertile land in the vicinity of Santiago and then devoted themselves to the task of gaining notice. Since they immigrated early and got rich quickly, they could claim the luxury of looking down on all those who came later. They married among themselves, and, being good Catholics, they produced a multitude of descendants. Their normal children were destined for the land, the ministry, and the church hierarchy, but never for commerce, which was reserved for a different class of people; the children who were less favored intellectually went into the navy. Often there was a son left over to become president of the republic. There are dynasties of presidents, as if the office were hereditary, because Chileans vote for a familiar name. The Errázuriz family, for example, provided three presidents, thirty-some senators, and I don't know how many politicians, besides several heads of the church. The virtuous daughters of "known" families married their cousins or became holy women who worked questionable miracles: unmanageable daughters were given to the care of the nuns. These families were conservative, devout, honorable, proud, and avaricious, though generally of good disposition—not so much by temperament as to assure winning favor in heaven. They lived in fear of God. I grew up convinced that every privilege comes as a natural consequence of a long list of responsibilities. That Chilean social class maintained a certain distance from lesser human beings because they had been placed on Earth to set an example, a heavy burden they assumed with Christian devotion. One thing I must make clear, however, is that despite their origins and their surnames, my grandfather's branch of the family was not of that oligarchy; they had good credentials but lacked land or fortune.

One of the characteristics of Chileans in general, and of the descendants of Spaniards and Basques in particular, is their seriousness, which contrasts with the exuberant temperament so common in the rest of Latin America. I grew up among millionaire aunts, cousins of my grandfather and my mother, who wore ankle-length black dresses and made a great virtue of "turning" their husbands' suits, a tedious process that consisted of ripping apart the suit, pressing out the pieces, and sewing them back together, inside out, to give them new life. It was easy to distinguish the victims of these labors because the breast pocket of their jacket was always on the right rather than the left. The result was consistently pathetic, but the effort demonstrated how thrifty and hardworking the wearer's good lady was. The tradition of industrious women is fundamental in my country, where sloth is a male privilege. It is forgivable in men, just as alcoholism is tolerated among them, because it is assumed that these are unavoidable biological characteristics: if you're born that way, you're born that way. . . . That isn't true of women, you understand. Chilean women, even those with fortunes, do not paint their fingernails, since that would indicate they don't work with their hands, and one of the worst possible epithets for a Chilean woman to be called is *lazy*. It used to be that when you got on a bus you would see all the women knitting; that's no longer true because now Chile is showered with tons of secondhand clothing from the United States and polyester garbage from Taiwan and knitting has passed into history.

There has been speculation that our ponderous seriousness is the bequest of exhausted Spanish conquistadors, who arrived half dead with hunger and thirst, driven more by desperation than by greed. Those valiant captains—the last to share in the booty of the conquest—had to cross the cordillera of the Andes through treacherous passes, slog across the Atacama Desert beneath a sun like burning lava, or defy the ominous seas and winds of Cape Horn. The reward was scarcely worth the trouble, because Chile, unlike other regions of the continent, did not offer the possibility of wealth beyond dreams. Gold and silver mines could be counted on the fingers of one hand, and the minerals had to be torn from the rock with unspeakable effort. Neither did Chile have the climate for prosperous tobacco, coffee, or cotton plantations. Ours has always been a country with one foot in the poorhouse; the most that the colonist could aspire to was a quiet life dedicated to agriculture.

Ostentation was once unacceptable, as I've said, but unfortunately that has changed, at least among the residents of Santiago. They have become so pretentious that they go to the supermarket on Sunday mornings, fill their carts with the most expensive items—caviar, champagne, filets mignons—walk through the store for a while so everyone can see what they're buying, then leave the cart in an aisle and slip out discreetly with empty hands. I've also heard that a good percentage of cell phones are made of wood, mere fakes to show off. Such behavior once would have been unthinkable. The only people who lived in mansions were nouveau riche Arabs, and no one in his right mind would have worn a fur coat, even if it was as cold as the South Pole.

The positive side of such modesty—false or authentic—was, of course, simplicity. None of those parties for fifteen-year-olds with pink-dyed swans, no imperial weddings with four-layer cakes, no parties, with orchestra, for lap dogs, as in other capital cities of our exuberant continent. Our national seriousness was a notable characteristic that disappeared with the advent of the all-out capitalism imposed in the last two decades, when to be rich and to show it became fashionable. The character of the people is deep-rooted, however. Ricardo Lagos, the current president of the republic (2002), lives with his family in a rented house in an unpretentious neighborhood. When dignitaries from other nations visit, they are startled by the small size of the house, and their amazement grows when they see the president prepare the drinks and the first lady help serve the table. Although the right does not forgive Lagos for not being "one of us," they admire his simplicity. This couple are typical exponents of the old middle class formed in free, humanist, state schools and universities. The Lagos are Chileans brought up with the values of equality and social justice, and today's materialistic obsession seems not to have rubbed off on them. It is hoped that their example will end once and for all the wooden cell phones and the shopping carts abandoned in supermarket aisles.

It occurs to me that this sobriety, so deeply rooted in my family, as well as our habit of veiling our happiness or well-being, was founded in the embarrassment we felt when we saw the poverty all around us. It seemed to us that having more than others did was not only divine injustice but also a form of personal sin. We had to do penance and practice charity to compensate. The penance was to eat beans, lentils, or chickpeas every day, and to freeze in the winter. The charity part was a routine family activity, which was almost exclusively the purview of women. From the time we were little girls, our mothers or aunts took us by the hand and led us out to distribute food and clothing to the poor. That custom ended some fifty years ago, but helping a neighbor is an obligation that Chileans happily assume today, as is only just in a country that has no lack of opportunity for doing good. In Chile, poverty and solidarity go hand in hand.

There is no doubt of a tremendous disparity between rich and poor, just as there is in nearly all of Latin America. At least the Chilean people, poor as they may be, are well educated, informed, and aware of their rights—though they don't always reap the benefits of them. Poverty, nevertheless, continues to raise its ugly head, especially in times of crisis. I can't resist the temptation to copy a paragraph my mother wrote me from Chile following the floods of the winter of 2002, which buried half the country in an ocean of filthy water and mud.

It's been raining for days. Suddenly it lets up and only a fine mist keeps everything wet. Just as the Ministry of the Interior says better weather is on the way, another downpour comes along and blows off your hat. This has been another trial for the poor. We've seen the true face of misery in Chile, poverty disguised as the lower middle class, those who suffer most because they have hopes. These people have worked a lifetime to get a decent place to live and it turns out they're swindled by builders: their homes are nicely painted on the exterior, but they don't have drainage, so the rain not only soaks them, the walls begin to crumble like stale bread. The only thing that distracts from the disaster is the world soccer championship. Iván Zamorano, our soccer idol, donated a ton of food and spends his days in flooded neighborhoods entertaining the kids and handing out soccer balls. You can't imagine the painful scenes; it's always those who have the least who suffer the worst misfortunes. The future looks black because this endless rain has all the vegetable fields under water and the wind has flattened whole orchards. In Magallanes sheep have died by the thousands, trapped in the snow at the mercy of the wolves. Of course Chilean solidarity can be seen everywhere. Men, women, and teenagers in water up to their knees and covered with mud are caring for children, handing out clothing, and shoring up entire neighborhoods that have been washed toward the ravines. They have set up an enormous tent in the Plaza Italia; cars drive by and without even stopping someone tosses out a bundle of blankets and food into the arms of a waiting student. The Mapocho station has been turned into an enormous shelter for the victims, and on the stage all of Santiago's artists, rock musicians, even the symphony orchestra, keep things lively, and people stiff with cold can't resist dancing and so for a few minutes forget their troubles. This has been an enormous lesson in humility. The president and his wife, along with all his ministers, are visiting the shelters and offering comfort. The greatest thing is that the minister of defense, Michelle Bachelet, the daughter of a man assassinated during the dictatorship, called out the army to come to the aid of those affected, and is touring in an armored car, with the commander in chief by her side, helping in every way possible, night and day. Everyone is doing what he or she can. The big question is what will the banks do, they're such a scandal in this country.

Just as a Chilean is annoyed by the success of others, he is equally magnanimous during disasters, at which time he sets aside his pettiness and is instantly converted into the most supportive and generous person in this world. There are several annual television marathons in Chile devoted to charity, and everyone, particularly the most humble among us, throws himself into a true frenzy to see who can give the most. Occasions for appealing to public compassion are never wanting in a nation eternally rocked by catastrophes that shake the foundations of life, floods that sweep away entire towns, gigantic waves that deposit ships in the center of a plaza. We are created in the idea that life is precarious, and we are always waiting for the next calamity to happen. My husband—who is six feet tall and a bit creaky in the knees—could never understand why I keep the glasses and plates on the lowest shelves in the kitchen, which he can reach only when lying on his back . . . until the 1988 earthquake in San Francisco destroyed the neighbors' china while ours escaped unscathed.

Not everything is guilt-ridden breast-beating and charitable works performed in order to redress economic injustices. Oh, no. Our seriousness is amply compensated by our gluttony; in Chile, life is lived around the dining-room table. Most of the executives I know suffer from diabetes because they hold their business meetings at breakfast, lunch, and dinner. No one signs a paper without indulging in, at least, cookies and coffee, or a drink.

While it is true that we eat beans every day, on Sundays the menu changes. A typical luncheon at my grandfather's house began with stick-to-the-ribs fried empanadas, meat pies with onion, which can provoke heartburn in the healthiest eater; then came a *cazuela,* a raise-the-dead soup of meat, corn, potatoes, and vegetables, followed by a succulent seafood *chupe* that flooded the house with its delicious aroma; and to end, we had a selection of irresistible desserts, which always included a tarte of *manjar blanco* or *dulce de leche,* a milk-based caramel (my aunt Cupertina's legendary recipe)—all accompanied by our fatal *pisco* sours and several bottles of good red wine that had been aged for years in the family cellar. Before we left, we were given a tablespoon of milk of magnesia. This dosage was increased by five when an adult birthday was being celebrated: we children didn't merit such deference. I never heard the word *cholesterol* mentioned. My parents, who are over eighty, consume ninety eggs, a quart of cream, a pound of butter, and four pounds of cheese per week. They're healthy and lively as little kids.

Family reunions were not only a fine opportunity for everyone to eat and drink till they dropped, but also a chance to do battle to the death. With the second *pisco* sour the screams and insults among my relatives could be heard through the whole barrio. Afterward, each person went his own way, swearing never to speak to the others again, but the next Sunday everyone was there: no one dared *not* come, my grandfather wouldn't have forgiven it. I understand that this pernicious custom is still being observed in Chile, even though there has been great progress in other regards. I always was intimidated by those compulsory reunions, but now, in the ripe years of my life, I have recreated them in California. My formula for an ideal weekend is to have the house filled with people, to cook for a regiment, and at the end of the day to hear everyone arguing at the tops of their voices.

Feuds among relatives were carried on in private. Privacy is a luxury of the well-to-do because most Chileans have none. Middle-class families and below live in very close quarters, in many homes several people sleep in the same bed. When there is more than one room, the dividing walls are so thin that every sigh comes right through. To make love you have to hide in unimaginable places: public baths, underneath bridges, at the zoo. Considering that the solution to the housing problem may take twenty years— and that's optimistic—it occurs to me that the government has the obligation to provide free motels for desperate couples. That way many mental problems could be avoided.

Every family has more than one troublemaker, but the modus operandi always is to close ranks around the black sheep and avoid scandal. From the cradle, we Chileans learn that "dirty linen is washed in private," and no one talks about alcoholic relatives, those with money problems, the ones who beat their wives or have served time. Everything is hidden, from a kleptomaniac aunt to the cousin who seduces little old ladies to relieve them of their pitiful savings, and particularly the distant male cousin who sings in a cabaret dressed as Liza Minnelli, because in Chile any originality in matters of sexual preference is unpardonable. There has been a real battle over discussing AIDS in public; no one wants to admit how it is transmitted. Neither is there legislation pertaining to abortion, one of the most serious health problems in the country; everyone hopes that if the subject isn't broached, it will disappear as if by magic.

My mother has a tape containing juicy anecdotes and family scandals, but she won't let me listen to it because she's afraid I'll divulge the contents. She has promised me that at her death, when she is absolutely safe from the apocalyptic vengeance of her blood kin, I will inherit that recording. I grew up surrounded with secrets, mysteries, whispers, prohibitions, matters that must never be mentioned. I owe a debt of gratitude to the countless skeletons hidden in our armoire because they planted the seeds of literature in my life. In every story I write I try to exorcise one of them.

In my family no one spread gossip, and in that we were somewhat different from the ordinary *homo chilensis* because the national sport is to talk about the person who just left the room. In this, too, we are different from our idols, the English, whose principles forbid them from making personal remarks. (I know a former soldier in the British army, married, the father of four children and grandfather of several, who decided to change gender. Overnight he appeared dressed as a woman and no one, absolutely no one, in the English village where he had lived for forty years made the least comment.) In Chile we even have a term for talking about our friends and neighbors—*plucking*—the etymology of which surely comes from plucking chickens, or denuding the out-of-earshot victim of his feathers. This habit is so prevalent that no one wants to be the first to leave, which is why farewells take an eternity at the door. In our family, in contrast, the norm of not speaking ill of others, a rule imposed by my grandfather, reached such an extreme that he never told my mother the reasons why he opposed her marriage to the man who would become my father. He refused to repeat the rumors that were circulating about his conduct and his character because he didn't have proof, and rather than defame my mother's suitor, he preferred to risk the future of his daughter, who in blissful ignorance ended up marrying a man who didn't deserve her. Over the years I have freed myself from this family trait. I have no scruples about repeating gossip, talking behind others' backs, or spreading their secrets in my books, the reason why half my relatives don't speak to me.

Relatives who don't speak to you are a common occurrence. The renowned novelist José Donoso found himself forced by family pressure to eliminate from his memoirs a chapter about an extraordinary great-grandmother, who when widowed opened a clandestine gaming house with attractive female croupiers. It's said that the stain on the family name prevented her son from becoming president, and a century later her descendants are still trying to hide her story. I regret that that great-granny didn't belong to my tribe. If she had, I would have taken justifiable pride in exploring her story. With an ancestor like her, think of all the delicious novels I could have written.

OF VICES AND VIRTUES

In my family nearly all the men studied law, although I don't remember a single one who passed the bar. The Chilean loves laws, the more complicated the better. Nothing fascinates us as much as red tape and multiple forms. When some minor negotiation seems simple, we immediately suspect that it's illegal. (I, for example, have always doubted that my marriage to Willie is valid, since it took place in fewer than five minutes, simply by writing our signatures in a book. In Chile it would have meant wading through several weeks of bureaucracy.) The Chilean is a legal animal. There's no better job in the country than being a notary public: we want everything on paper, sealed, with multiple copies and stamps on every page. We are so legalistic that General Pinochet wanted to pass into history as a president, not a usurper of power, and to do that he had to change the constitution. Through one of those ironies that are so abundant in history, he later found himself trapped in the laws he himself had created in order to perpetuate his tenure in office. According to the terms of his constitution, he was to fulfill his role for eight additional years—he had already been in power for several—that is, until 1988, when he would call a referendum so the people could decide whether he was to continue or to call an election. He lost that referendum and the following year lost the election and had to turn over the presidential sash to his opponent, the democratic candidate. It's difficult to explain to anyone outside of Chile how the dictatorship could be brought down when it could count on the unconditional support of the armed forces, the right, and a large part of the population. Political parties had been suspended, Congress disbanded, and the press was censored. As the general had often boasted: "Not a leaf stirs in this country without my consent." How, then, could he have been defeated in a democratic vote? This could happen only in Chile. In similar fashion, using a loop-hole in the law, an attempt is currently underway to try him along with other military men, even though he had appointed the Supreme Court and an amnesty protects the military from bearing responsibility for illegal acts committed during the years of his government. It turns out that there are hundreds of persons who were arrested but whom the military denies having killed; since they haven't appeared, it is assumed that they were kidnapped. In such cases the crime remains on record, so the accused cannot take shelter behind the amnesty.

Love for regulations, however unworkable they may be, finds its best exponents in the enormous bureaucracy of our suffering country. That bureaucracy is the paradise of the people in their uniform gray suits. There such a person can vegetate to his pleasure, completely safe from the traps of imagination, perfectly secure in his post to the day he retires—unless he is imprudent enough to try to change things, an observation made by the author-sociologist Pablo Huneeus (who is, I might add in passing, one of the few eccentric Chileans who isn't related to my family). A public official must understand from his first day in office that any show of initiative will signal the end of his career because he isn't there to be meritorious but to reach his level of incompetence with dignity. The point of moving papers with seals and stamps from one perusal to the next is not to resolve problems, but to obstruct solutions. If the problems were resolved, the bureaucracy would lose power and many honest people would be left without employment; on the other hand, if things get worse, the state increases the budget and hires more people, and thus lowers the index of the unemployed. Everyone is happy. The official abuses every smidgen of his power, starting from the premise that the public is his enemy, a sentiment that is fully reciprocated. It was a shock to find that in the United States all that's needed to move about the country is a driver's license, and that most transactions can be accomplished by mail. In Chile, the clerk on duty demands that the poor petitioner produce proof that he was born, that he isn't a criminal, that he paid his taxes, that he registered to vote, and that he's still alive, because even if he throws a tantrum to prove that he hasn't died, he is obliged to present a "certificate of survival." The problem has reached such proportions that the government itself has created an office to combat bureaucracy. Citizens may now complain of being shabbily treated and may file charges against incompetent officials . . . on a form requiring a seal and three copies, of course. Recently, a busload of us tourists crossing the border between Chile and Argentina had to wait an hour and a half while our documents were checked. Getting through the Berlin Wall was easier. Kafka was Chilean.

I believe that this obsession of ours with legality is a kind of safeguard against the aggression we carry inside; without the nightstick of the law we would go after one another tooth and claw. Experience has taught us that when we lose control we are capable of the worst barbarism, and for that reason we try to move cautiously, barricading ourselves behind bulwarks of paper bearing seals. Whenever possible, we try to avoid confrontation; we seek a consensus and, at the first opportunity, we put any decision to a vote. We love to vote. If a dozen kids get together in the schoolyard to play soccer, the first thing they do is write a set of rules and vote for a president, a board of directors, and a treasurer. This doesn't mean that we're tolerant, far from it: we cling to our ideas like maniacs (I am a typical case). You see intolerance everywhere, in religion, in politics, in the culture. Anyone who dares dissent is squelched with insults or ridicule, in the event that he can't be made to shut up using more drastic methods.

In customs we are conservatives and traditionalists; we prefer the known evil to good yet to be learned, but in everything else we are always on the lookout for something new. We have the idea that anything that comes to us from outside the country is by nature better than ours and should be tried, from the latest electronic gadget to economic and political systems. We spent a good part of the twentieth century trying out various forms of revolution, from Marxism to savage capitalism, ranging through each and every intermediate shading. Our hope that a change in government can improve our luck is like hoping to win the lottery: totally without rational foundation. At heart we know very well that life isn't easy. Ours is a land of earthquakes, why wouldn't we be fatalists? Given the circumstances, we have no choice but to be also a little stoic—though there's no reason to be too dignified about it; we are free to complain all we want.

In my family's case, I believe we were easily as spartan as stoic. According to my grandfather, cancer is caused by easy living, whereas discomfort is good for the health. He recommended cold showers, food difficult to chew, lumpy mattresses, third-class seats on trains, and clunky shoes. His theory of healthful discomfort was reinforced by several English schools for young girls in which it was my destiny to spend the greater part of my childhood. If you survive this kind of education, you are forever after grateful for the most trivial pleasures. I'm a person who murmurs a silent prayer of thanks when warm water comes out of the tap. I expect life to be problematic, and when for several days I haven't suffered anguish or pain, I get worried because I am sure that means the heavens are preparing a worse misfortune for me. Even so, I am not totally neurotic, quite the opposite; in fact, I'm quite easy to be with. . . . It doesn't take much to make me happy; you know now that ordinary warm water in the faucet will do the trick.

It has often been said that we Chileans are envious, that we are bothered by others' success. It's true, but the explanation is that what we're feeling isn't envy, it's common sense. Success isn't normal. The human being is biologically constituted for failure, the proof of which being that we have legs instead of wheels, elbows instead of wings, and metabolism instead of batteries. Why dream of success if we can calmly vegetate in our failures? Why do today what we can put off till tomorrow? Or do well what we can do halfway? We detest it when a countryman rises above the rest of us, except when it happens in another country, in which case the lucky fellow (or female equivalent) becomes a kind of national hero. The person who triumphs locally, however, is less than adored; soon there is tacit accord that he should be taken down a peg or two. We call this sport *chaqueteo,* "jacketing": grabbing the offender by his coattails and pulling him down. Despite the *chaqueteo* and an ambience of mediocrity, from time to time someone does manage to raise his head above the crowd. Our country has produced exceptional men and women: two Nobel laureates—Pablo Neruda and Gabriela Mistral—the singers/composers Víctor Jara and Violeta Parra, the pianist Claudio Arrau, the painter Roberto Matta, and the novelist José Donoso, to mention a few who come to mind.

We Chileans enjoy funerals because the dead person is no longer a rival, and now he can't backstab us. Not only do we go to burials en masse, where we have to stand for hours listening to at least fifteen speeches, we also celebrate the anniversaries of the deceased's death. One of our entertainments is telling and listening to stories, the more macabre and tragic the better; in that, and in our taste for "the drink," we resemble the Irish. We're addicts of radio and television soaps; the misadventures of the protagonists offer us a good excuse to weep for our own sorrows. I grew up listening to dramas in the kitchen, despite my grandfather's having outlawed the radio because he considered it an instrument of the devil used to propagate gossip and vulgarity. We children and servants suffered through the endless ordeals of the serial *Right to Live* for several years, as I remember.

The lives of the characters in TV soaps are much more important than those of family, even though the plot isn't always easy to follow. For example, the handsome lead seduces a woman and leaves her in an interesting condition, then for revenge he marries a girl who is lame, and leaves her, too, "baby waiting," as we say in Chile, but right away he runs off to Italy to join his first wife. I think this is called trigamy. In the meantime, the second girl has her lame leg operated on, goes to the beauty shop, inherits a fortune, becomes an executive in a large corporation, and attracts new suitors. When the lead returns from Italy and sees that rich young female with two legs the same length, he repents of his felony. And then begin the writer's problems in untangling the mare's nest the story has become. The first seduced woman gets an abortion, so there won't be any bastards running around that TV channel, and he kills off the luckless Italian, thus the lead—who apparently is the good guy in the series—is opportunely widowed. This allows the formerly lame heroine to marry in white, despite the enormous belly she's sporting, and within a short time she gives birth to . . . a baby boy, of course. No one works; they live on their passions, and the women all go around wearing false eyelashes and cocktail dresses from early morning on. In the course of this tragedy nearly all the actors end up hospitalized; there are births, accidents, rapes, drugs, teenagers who run away from home or from prison, blind men, madmen, rich who become poor and poor who become rich. There's a lot of suffering. The day after a particularly dramatic chapter, telephones all across the country are buzzing with the details: my friends call collect from Santiago to California to keep me up to date. The only thing that can compete with the last chapter of a TV series is a visit from the pope, and that has happened only one time in our history, and likely won't happen again.

Besides funerals, morbid stories, and soaps, we count on crimes, always an interesting subject for conversation. We are fascinated with psychopaths and murderers; if they're upper class, so much the better. "We have a bad memory for crimes of state, but we never forget the peccadilloes of the man next door," commented a famous journalist. One of the best-known murders in our history was committed by a Señor Barceló, who killed his wife after having abused her all during the years they shared together and then alleged it had been an accident. I was embracing her, he said, and my gun misfired and a bullet penetrated her brain. He couldn't explain why he had a loaded pistol in his hand, pointed at the nape of her neck. Given that information, his mother-in-law initiated a crusade to avenge her unfortunate daughter; I don't blame her, I would have done the same. This woman came from the highest level of Santiago society, and was used to getting her way. She published a book denouncing her son-in-law, and after he was sentenced to death she installed herself in the office of the president of the republic to prevent him from granting a pardon. The villain was executed. He was the first, and one of the few, upper-class prisoners to receive the death penalty because that punishment was reserved for those who had no connections or good lawyers. Today the death penalty has been eliminated, as it has in any decent country.

I also grew up with the family anecdotes told to me by my grandparents, my uncles, and my mother—very handy when it comes to writing novels. How many of them are true? Doesn't matter. At the hour of remembering, no one wants verification of facts, the legend is enough, like the sad tale of the ghost that during a séance told my grandmother the location of a treasure hidden beneath the stairs. Due to an error in the plan of the property, not because of any malevolence on the part of the spirit, the treasure was never found, even though they tore down half the house. I've tried to verify the how and the when of these lamentable events, but no one in my family is interested in documenting them, and if I ask a lot of questions, my relatives get offended.

I don't want to give the impression that we are all bad, we also have our virtues. Let me see, I'll try to think of one. . . . Well, we're a people with poetic souls. It isn't our fault; that one we can blame on the landscape. No one who is born and lives in a natural world like ours can resist writing poetry. In Chile, you lift up a rock, and instead of a lizard out crawls a poet or a balladeer. We admire our writers, we respect them, and we put up with their manias. Years ago at political meetings people would shout aloud the poems of Pablo Neruda, which we all knew by heart. We liked his love poems best because we have a weakness for romance. We are also moved by misfortune: dejection, nostalgia, disillusion, grief. Our evenings are very long, which may explain our preference for melancholy themes. If poetry isn't your thing, there are always other forms of art. All the women I know write, paint, sculpt, or do crafts in their leisure time—which is very scarce. Art has replaced knitting. I've been given so many paintings and ceramics that I can no longer get my car in the garage.

I can add about our character that we're affectionate; we go around bestowing kisses right and left. We greet each other with a sincere kiss on the right cheek. Children kiss adults as they arrive and as they leave, and as an additional sign of respect they call them *uncle* or *aunt,* as they do in China, and that includes schoolteachers. Older people are kissed mercilessly, even against their will. Women kiss, even if they hate each other, and they kiss any male within reach, and neither age nor social class nor hygiene can dissuade them. Only males in their reproductive years, let's say between fourteen and seventy, do not kiss each other—with the exception of father and son—but they clap each other on the back and heartily embrace. This affection has many other manifestations, from opening the doors of your house to receive anyone who shows up unexpectedly to sharing everything you have. It never crosses your mind to praise something another person is wearing, because they're certain to whip it off and give it to you. If there is food left from a meal, the genteel thing is to give it to the guests to take home, just as you never arrive at someone's house with empty hands.

The first thing you can say about Chileans is that we are friendly and hospitable; at the first hint we throw open our arms and the doors of our homes. I've often heard foreigners say that if they ask directions, the people they approach accompany them there personally, and if they seem to be lost, their informant is capable of inviting them home for dinner, even offering a bed if they're in difficulty. I confess, however, that my own family was not especially friendly. One of my uncles would not allow anyone to breathe near him, and my grandfather was given to thrashing the telephone with his cane because he considered it a lack of respect to call without his consent. He was perpetually cross with the mailman because he brought unsolicited mail, and he never opened letters unless the sender's name was prominently displayed. My relatives felt they were superior to the rest of humankind, though their rationale always seemed nebulous to me. According to my grandfather's school of thought, we could trust no one but close family; the rest of humanity was suspect. My grandfather was a fervent Catholic but he deplored confession; he was suspicious of priests and he believed that forgiveness of sins could be negotiated directly with God, just as he could negotiate for those of his wife and his children. Despite this inexplicable superiority complex, visitors were always warmly received in our home, however vile they might be. In this sense, we Chileans are like the Arabs of the desert: the guest is sacred, and friendship, once declared, is an indissoluble bond.

It is impossible to go into a home, rich or poor, without accepting something to eat or drink, even if it's only a cup of tea. This is another national tradition. Since coffee has always been scarce, and expensive—even Nescafé was a luxury—we drank more tea than the entire population of Asia put together, but on my last trip I found to my amazement that coffee finally had made its entrance into the culture, and now anyone willing to pay can find espressos and cappuccinos worthy of Italy. In passing I should add, for the peace of mind of potential tourists, that impeccable public bathrooms and bottled water are readily available everywhere; it's no longer mandatory to come down with colitis after your first glass of water, as it once was. In a strange way, I lament that, because those of us who grew up drinking Chilean water are immunized against all known and yet-to-be-discovered bacteria. I can drink water from the Ganges with no visible effect on my health; my husband, in contrast, once outside the United States, can brush his teeth with bottled water and still contract typhus. In Chile we are not refined in respect to tea, any brew sweetened with a spot of sugar tastes delicious to us. There are, in addition, an infinite number of local herbs to which we attribute curative properties, and in the case of the truly poor, we have *agüita perra*—bitch water—nothing but plain hot water in a cracked teacup. The first thing we offer a visitor is a *tecito,* an *agüita,* or a *vinito,* a "nice little drink" of tea, water, or wine. We always add the diminutive *-ito* to our words, almost as an apology for offering, in accord with our desire not to be noticed and our horror of putting on airs, even with words. Then we offer our guest "pot luck," which means that the mistress of the house will take bread out of her children's mouths to give to the visitor, who is obliged to accept it. If you receive a formal invitation, you can expect a gargantuan feast:

the goal is to leave the guests moaning with indigestion for several days. Of course, women always do the hard work. Now it's considered chic for men to cook, a really bad development because while they take all the glory, the woman has to wash up the mounds of pots and dirty dishes he's left everywhere. Our typical cuisine is simple because earth and sea are generous; there is no fruit or seafood more delicious than ours—that I can assure you. The more difficult it is to put food on the table, the more elaborate and spicy it becomes, witness the examples of India and Mexico, where there are three hundred ways to cook rice. We have one, and that seems more than sufficient to us. We don't need to be creative or invent original dishes, we do that with names, which can lead the foreigner to the worst suspicions: "beat-up fools" (an abalone dish), "head cheese," "dark blood," "fried brains," "lady fingers," "queen's arm," "nun's sighs," "wrapped babies," "torn bloomers," and "monkey tail," for a starter.

We Chileans have a sense of humor and we like to laugh, even though deep down we prefer seriousness. We had a president named Jorge Alessandri (1958–1964) who was a neurotic bachelor; he drank only mineral water and never allowed anyone to smoke in his presence, and people always said about him, with admiration, "How sad our Don Jorge is!" That calmed us because it was a sign that we were in good hands, those of a serious man, or, better still, of an aging depressive who wasted no time on pointless happiness. Which is not to say that we don't find bad luck entertaining;

we sharpen our sense of humor when things go badly, and since it seems things always go badly, we laugh a lot. That's a small compensation for our vocation of complaining about everything. A person's popularity is measured by the number of jokes about him. They say that President Salvador Allende invented jokes about himself—some more than a little racy—and set them loose on the world. For many years I had a magazine column and a television program with humorous pretensions, which were tolerated because there was very little competition—in Chile even clowns are melancholy. Years later, when I began to publish a similar column for a newspaper in Venezuela, it bombed and brought me a mountain of enemies besides, because humor in that country is more direct and not as cruel.

My family is famous for practical jokes, but it may lack taste in matters of humor; the only jokes my relatives understand are German stories about Herr Otto. Here's one example: A very elegant woman broke wind, involuntarily and loudly, and to cover it up made a noise with her shoes. Then Herr Otto says (it has to be in a German accent), "You can break a shoe, you can break a heart, but you'll never make the noise you made with that fart." As I'm writing this, I'm weeping with laughter. I've tried to tell the joke to my husband, but it doesn't translate, and besides, in California ethnic jokes are not at all in favor. I grew up with jokes about Galicians, Jews, and Turks. Our humor is black. We never let an opportunity pass to make fun of other people, whoever they may be: deaf mutes, the retarded, epileptics, people of color, homosexuals, priests, and the homeless. We have jokes about all religions and races. The first time I heard the expression "politically correct" I was forty-five years old, and I have never been able to explain to friends or relatives in Chile what that means. Once in California I tried to get one of those dogs they train to lead the blind but are given away when they can't pass the rigorous tests. In my application I had the bad idea of mentioning that I wanted a "rejected" dog, and by return mail I received a dry note informing me that the term "rejected" is never used; instead, you say that the animal "has changed careers." Try and explain that in Chile!

My mixed marriage with a gringo hasn't gone all that badly; we get along (even though most of the time neither of us has the least idea of what the other is talking about) because we are always ready to give each other the benefit of the doubt. The greatest drawback is that we don't share a sense of humor. Willie can't believe that I can be funny in Spanish, and as for me, I never know what the devil he's laughing about. The one thing that amuses us both at the same time are the off-the-cuff speeches of President George W. Bush.

THE ROOTS OF NOSTALGIA

I have often said that my nostalgia dates from the time of the military coup of 1973, when my country changed so much that I can no longer recognize it, but in fact it must have begun much earlier. My childhood and adolescence were marked with journeys and farewells. I hadn't yet put down roots in one place when it was time to pack our suitcases and move to another.

I was nine years old when I left my childhood home and with great sadness said good-bye to my unforgettable grandfather. So I would be entertained during my trip to Bolivia, Tío Ramón gave me a map of the world and the complete works of Shakespeare in Spanish, which I swallowed at a gulp, reread several times, and still own. I was fascinated by those stories of jealous husbands who murdered their wives over a handkerchief, kings whose enemies put distilled poison in their ears, lovers who committed suicide because of faulty communication. (How different Romeo and Juliet's fate would have been if they'd had a telephone!) Shakespeare initiated me into stories of blood and passion, a dangerous road for authors like me whose fate it is to live in a minimalist era. The day that we set out for the province of Antofagasta, where we were to take a train to La Paz, my mother gave me a notebook and instructed me to start a travel diary. Ever since then I have written almost every day; writing is my most deeply entrenched habit. As that train chugged across the countryside, the landscape changed and I felt something tear inside me. On the one hand I was curious about all the new things passing before my eyes, and on the other, an insurmountable sadness was crystallizing deep within me. In the small Bolivian towns where the train stopped, we bought corn on the cob, leavened bread, black potatoes that looked rotten, and delicious sweets, all offered by Bolivian Indian women in multicolored wool skirts and black derbies like those worn by English bankers. I wrote everything down in my notebook with the industry of a notary, as if even then I foresaw that only writing would anchor me to reality. Outside the window, the world was hazy because of the dust on the glass, and deformed by the speed of the train.

Those days shook my imagination. I heard stories of the spirits and demons that wander the abandoned towns, of mummies exhumed from profaned tombs, of hills of human skulls, some more than fifty thousand years old, exhibited in a museum. In school, in history class, I had learned that the first Spaniards to reach Chile in the sixteenth century, coming from Peru, had wandered for months through these desolate reaches. I imagined that handful of warriors in red-hot armor, their exhausted horses, their hallucinated eyes, followed by a thousand captive Indians carrying provisions and weapons. It was a feat of incalculable courage and mad ambition. My mother read us some pages about the now vanished Atacameño Indians, and about the Quechuas and Aymaras, among whom we would live in Bolivia. Although I couldn't know that yet, my destiny as a vagabond began on that journey. That diary still exists today; my son has hidden it and refuses to show it to me, because he knows I would destroy it. I regret many things I wrote in my youth: frightful poems, tragic stories, suicide notes, love letters addressed to unfortunate lovers, and especially that dreadful diary. (A caution to aspiring writers: not everything you write is worth keeping for the benefit of future generations.) When she gave me that notebook, my mother somehow intuited that I would have to dig up my Chilean roots, and that lacking a land into which to sink them I would have to do that on paper. I maintained a correspondence with my grandfather, my Tío Pablo, and with the parents of some friends, patient people to whom I related my impressions of La Paz, its purple mountains, its hermetic Indians, and its air, so thin that your lungs are always on the verge of filling with foam and your mind with hallucinations. I didn't write to children my own age, only adults, because they answered my letters.

In my childhood and youth, I lived in Bolivia, the Middle East, and Europe, following the diplomatic destiny of the "dark, mustached man" the gypsies foretold so many times. I learned a little French and English, and also learned to eat suspicious-looking food without asking questions. My education was chaotic, to say the least, but I compensated for enormous gaps in information by reading everything that fell into my hands with the voraciousness of a piranha. I traveled by ship, plane, train, and automobile, always writing letters in which I compared what I saw with my one eternal reference: Chile. I never left behind the flashlight my Tío Pablo gave me, which helped me read in the most adverse conditions, or the notebook that contained the story of my life.

After two years in La Paz, we set off with bag and baggage for Lebanon. The three years in Beirut were a time of isolation for me, confined as I was to my home and my school. How I missed Chile! At an age when girls were dancing to rock 'n' roll, I was reading and writing letters. Elvis Presley was already fat by the time I learned of his existence. I wore depressing gray dresses to annoy my mother who was always elegant and attractively dressed, while at the same time I daydreamed of princes fallen from the stars who would rescue me from a banal life. During recess in school, I would barricade myself behind a book in the farthest corner of the schoolyard, to disguise my shyness.

The adventure in Lebanon ended abruptly in 1958, when the U.S. Marines of the Sixth Fleet disembarked to intervene in the violent political squabbles that soon would tear that country apart. Their civil war had begun months earlier amid sounds of gunfire and shouting; there was confusion in the streets and fear in the air. The city was divided into religious sectors that clashed over grudges accumulated through centuries while the army tried to keep order. One by one the schools closed their doors . . . all except mine because our phlegmatic director decided that since Great Britain wasn't involved, the war was none of her concern. Unfortunately, this interesting situation was short-lived: Tío Ramón, frightened by the direction the conflict was taking, sent my mother to Spain with the dog, and us children back to Chile. Later he and my mother were dispatched to Turkey, but we stayed in Santiago, my brothers in a boarding school and I with my grandfather.

I was fifteen when I returned to Santiago, disoriented from having lived several years outside the country and from having lost my ties with my old friends and my cousins. I talked with a strange accent to boot, which is a problem in Chile, where people are "situated" within social classes by the way they speak. Santiago at the time seemed very provincial to me compared, for example, with the splendor of Beirut, which boasted of being the Paris of the Middle East. That didn't mean that the rhythm of life was calm, not in the least, for Santiaguinos were already suffering from frayed nerves. Life was uncomfortable and difficult, the bureaucracy crushing, the working hours very long, but I arrived there determined to adopt that city in my heart. I was tired of telling people and places good-bye, I wanted to put down roots and never leave. I think I fell in love with my country because of the stories my grandfather told me and because of our travels together through the south. He taught me history and geography, showed me maps, made me read Chilean writers, corrected my grammar and handwriting. As a teacher, he was short on patience but long on severity; my errors made him red with anger, but if he was content with my work he would reward me with a wedge of Camembert cheese, which he ripened in his armoire; whenever he opened that door, the odor of stinking army boots flooded the neighborhood.

My grandfather and I got along well because we both liked sitting without talking. We could spend hours that way, side by side, reading or watching the rain drum against the windowpanes, without feeling any need for small talk. I believe we had a mutual liking and respect for one another.

I write that word, *respect,* with some hesitation because my grandfather was authoritarian and *machista;* he was used to treating women like delicate flowers, but the idea of any intellectual respect for them never crossed his mind. I was a prickly, rebellious fifteen-year-old girl who argued with him as equal to equal. That piqued his curiosity. He would smile with amusement when I claimed the right to the same freedom and education as my brothers, but at least he listened. It's worth mentioning here that the first time my grandfather heard the word *machista* it came from my lips. He didn't know what it meant, and when I explained, he nearly died laughing; the idea that male authority, as natural as the air he breathed, had a name seemed naïve and laughable. When I began to question that authority, he didn't find it funny anymore, but I think he understood and perhaps admired my desire to be like him, strong and independent and not the victim of circumstances, as my mother had been.

I nearly succeeded in being like my grandfather, but nature betrayed me when one day two little cherries popped out on my ribs and my plan went all to hell. That hormone explosion was a disaster for me. In a matter of weeks, I was transformed into a complex-ridden girl whose head was swimming with romantic dreams, her sole preoccupation being how to attract the opposite sex—not an easy task since I hadn't an ounce of charm and was always in a rage. I couldn't veil my scorn for most boys because it was obvious to me that I was cleverer than they were. (It took me a couple of years to learn how to play dumb so that men would feel superior. You can't imagine the effort that takes!) I went through those years torn between the feminist ideas fermenting in my head—though incapable of articulating them in an intelligible way since no one in my world had ever heard such ideas expressed—and the longing to be like the rest of the girls my age, to be accepted, desired, conquered, protected.

It fell to my poor grandfather to cross swords with the most miserable adolescent in the history of humankind. Nothing the poor man said could console me. Not that he said much. Sometimes he muttered that I wasn't bad for a woman, but that didn't change the fact that he would have preferred me to be a man, in which case he would have taught me to use his tools. At least he managed to get rid of my gray, severely tailored dress through the simple expedient of burning it in the patio. That little caper sent me into a tantrum, but deep down I was grateful, even though I was sure that with or without that gray rag no man would ever look at me. A few days later, however, a miracle happened: Miguel Frías, my first boyfriend, asked me to be his girlfriend. I was so desperate that I latched onto him like a crab and never let go. Five years later we were married; we had two children and stayed together twenty-five years. But I don't want to get ahead of my story . . .

By that time my grandfather had retired his mourning clothes and married a matron with imperial bearing, no doubt the blood of those German colonists who during the nineteenth century had left the Black Forest to populate the south of Chile. In comparison to her we seemed like savages, and we behaved like them as well. My grandfather's second wife was an imposing Valkyrie, tall, white-skinned, blond, gifted with a magnificent prow and a memorable stern. She had to put up with a husband who murmured his first wife's name in his sleep and do battle with her in-laws, who never completely accepted her and on more than one occasion made her life impossible. I regret that now; without her the patriarch's last years would have been very lonely. She was an excellent cook and mistress of the house; she was also bossy, hardworking, thrifty, and at a loss to understand our family's twisted sense of humor. During her reign the eternal beans, lentils, and chickpeas were banished; she cooked delicate dishes that her stepsons doused with hot sauce even before tasting them, and she embroidered lovely towels they often used to scrub the mud from their shoes. I imagine that Sunday luncheons with those barbarians must have been an insufferable torment for her, but she continued them for decades to demonstrate to us that we would never defeat her, no matter what we did. In that battle of wills, she won by a mile.

This dignified lady was never included in the time I shared with my grandfather, but she sat with us at night, knitting by memory as we listened to a horror story on the radio with the lights out, she indifferent to the program and my grandfather and I nearly ill from terror and laughing. He had reconciled his differences with the media by then, and had an antediluvian radio that he spent half the day repairing. With the help of a *maestro*, he had installed an antenna and some cables connected to a metal grille, hoping to capture communications from extraterrestrials since my grandmother wasn't at hand to summon them in her sessions.

In Chile we have the institution of the *maestro*, as we call anyone (though never a woman) who has a pair of pliers and some wire in his power. If this person is especially primitive in his approach, we affectionately call him a *maestro chasquilla*, that is, maybe only a little scruffy; otherwise he was plain *maestro*, an honorary title equivalent to *licenciado*, our designation for almost anyone who has graduated from college. With pliers and some wire, this fellow can fix anything from a lavatory to an airplane turbine: his creativity and daring are boundless. Through the greater part of his long life, my grandfather rarely needed to call on one of these specialists, because not only was he able to correct any imperfection, he also fabricated his own tools. In his later years, however, when he couldn't bend down or lift a heavy weight, he counted on a *maestro*, who came to work with him . . . between slugs of gin. In the United States, where workmen are expensive, half the male population has a garage filled with tools and learns at a young age to read instruction manuals. My husband, a lawyer by profession, owns a pistol that shoots nails, a machine for cutting rock, and another that vomits cement through a hose. My grandfather was an exception among Chileans because no man from the middle class up knows how to decipher a manual, nor does he dirty his hands with motor oil—that's what *maestros* are for; they can improvise ingenious solutions with the most modest resources and a minimum of fuss. I knew one who fell from the ninth floor while trying to repair a window, and miraculously emerged without injury. He went back up in the elevator, rubbing his bruises, to apologize for having broken the hammer. The idea of using a safety belt or filing for compensation never entered his mind.

There was a little hut at the back of my grandfather's garden, surely built for a maid, and I made myself a nest there. For the first time in my life I had privacy and silence, a luxury to which I became addicted. I studied during the day and at night I read the sci-fi novels I rented for a few pennies at a nearby kiosk. Like all teenaged Chileans then, I walked around with *The Magic Mountain* and *Steppenwolf* under my arm to impress everyone, but I don't remember ever having read them. (Chile is possibly the one country where Thomas Mann and Hermann Hesse have been permanent best-sellers, although I can't imagine that we have anything in common with characters like Narcissus and Goldmund.) In my grandfather's library, I came across a collection of Russian novels and the complete works of Henri Troyat, who wrote long family sagas about life in Russia before and during the Revolution. I read and reread those books, and years later I named my son Nicolás after one of Troyat's characters, a young country man, radiant as a sunny morning, who falls in love with his master's wife and sacrifices his life for her. The story is so romantic that even today, when I think of it, it makes me want to weep. That's how all my favorite books were, and still are: passionate characters, noble causes, daring acts of bravery, idealism, adventure, and, when possible, distant locales with terrible climates, like Siberia or some African desert, that is, somewhere I never plan to go. Tropical islands, so pleasant for vacations, are a disaster in literature.

I was also writing to my mother in Turkey every day. Letters took two months to be delivered, but that has never been a problem for us, we have the vice of epistolary communication: we have written nearly every day for forty-five years, with the mutual promise that when either of us dies, the other will tear up the mountain of accumulated letters. Without that guarantee we couldn't write so freely. I don't want to think of the uproar that would result if those letters, in which we have made mincemeat of our relatives and the rest of the world, fell into indiscreet hands.

I remember those winters in my adolescence, when the rain engulfed the patio and flowed beneath the door of my little hut, when the wind threatened to carry off the roof, and thunder and lightning rattled the world. If I had been able to stay closed up there reading all winter, my life would have been perfect, but I had to go to class. I despised waiting for the bus, tired and anxious, never knowing whether I would be one of the fortunate who got on or one of the poor wretches who didn't make it and had to wait for the next bus. The city had spread out and it was difficult to get from one point to another; to get onto a bus (a *micro* to us) was tantamount to a suicide mission. After waiting hours along with twenty citizens as desperate as you, sometimes in the rain, standing ankle-deep in a mud pit, you had to run like a rabbit as the vehicle approached, coughing and belching smoke from the exhaust, and hop and grab a handhold on the steps, or on some passenger who'd been lucky enough to get his foot in the door. Not too surprisingly, this has changed. Today the *micros* are quick, modern, and numerous. The one drawback is that their drivers compete to be first at the bus stop in order to collect the maximum number of passengers, so they fly through the streets flattening anything in their way. They detest schoolchildren because they pay less, and old people because they take so long to get on and off, so they do anything in their power to prevent them from getting within a mile of their vehicle. Anyone who wants to know a Chilean's true character must use public transportation in Santiago and travel across the country by bus: the experience is most instructive. Street minstrels get onto the buses, magicians, jugglers, thieves, lunatics, and beggars, along with people selling needles, calendars, and color prints of saints and flowers. In general, Chileans are bad-humored and in the street never look you in the eye, but on the *micros* a kind of solidarity is established that resembles the camaraderie in London's air raid shelters during the Second World War.

One further word about traffic: Chileans, so timid and amiable in person, become savages when they have a steering wheel in their hands; they race to see who can be first to reach the next red light, they snake in and out of lanes without signaling, shout insults or make obscene gestures. Nearly all our epithets end in *-ón,* which makes them sound like French. A hand held out as if begging for alms is a direct allusion to the size of the enemy's genitals; that's good to know before you're foolish enough to place a coin in the offending palm.

With my grandfather I made some unforgettable trips to the coast, the mountains, and the desert. He took me twice to sheep ranches in the Argentine Patagonia, true odysseys by train, jeep, ox cart, and horseback. We traveled to the south through magnificent forests of native trees, where it is always raining; we sailed the pure waters of lakes that mirrored snowy volcanoes; we crossed through the rugged cordillera of the Andes along hidden routes used by smugglers. Once on the other side, we were met by Argentine herders, crude, silent men with able hands and faces tanned like the leather of their boots. We camped beneath the stars, wrapped in heavy wool ponchos, using our saddles for pillows. The herdsmen killed a kid and roasted it on a spit; we ate it washed down with *mate,* a green, bitter tea served in a gourd passed from hand to hand, all of us sipping from the same metal straw. It would have been an insult to have turned up my nose at a tube slick with saliva and chewing tobacco. My grandfather never believed in germs, for the same reason he didn't believe in ghosts: he'd never seen one. At dawn we washed in frosty water and strong yellow soap made from sheep fat and lye. Those journeys left me with such an indelible memory that thirty-five years later, when I told the story of the flight of the protagonists of my second novel, *Of Love and Shadows,* I could describe the experience and the landscape without hesitation.

CONFUSED YEARS OF YOUTH

During my childhood and youth, I saw my mother as a victim, and decided early on that I didn't want to follow in her footsteps. I was a feminist long before I'd heard the word; my need to be independent and not to be controlled by anyone is so old that I can't remember a single moment when it didn't guide my decisions. When I look back at the past, I realize that my mother was dealt a difficult destiny and in fact confronted it with great bravery, but at the time I judged her as being weak because she was dependent on the men around her, like her father and her brother Pablo, who controlled the money and gave the orders. The only time they paid any attention to her was when she was ill, so she often was. Later she began her life with Tío Ramón, a man of magnificent qualities but one who was at least as macho as my grandfather, my uncles, and the rest of Chilean manhood in general.

I felt asphyxiated, a prisoner in a rigid system—we all were, particularly the women around me. I couldn't take a step outside the norms; I had to be like all the others, sink into anonymity or encounter ridicule. It was assumed that I would graduate from high school, keep my sweetheart on a short rein, marry before I was twenty-five—any later and all was lost—and rapidly produce children so no one would think I used contraceptives. And in regard to that, I should clarify that the famous pill responsible for the sexual revolution had already been invented, but in Chile it was spoken of only in whispers; it was forbidden by the Church and could be acquired only through a physician friend of liberal inclinations . . . after producing a marriage license, naturally. Unmarried women were out of luck, because few Chilean men, even today, are civil enough to use a condom. In tourist guides they should recommend that visitors always carry one in their billfolds because they won't lack for opportunity to use them. For a Chilean, the seduction of any woman in her reproductive years is a conscientiously executed task. Although usually my compatriots are terrible dancers, they are accomplished sweet-talkers; they were the first to discover that a woman's G spot is in her ears, and that to look for it any lower is a waste of time. One of the most therapeutic experiences for any depressed woman is to walk past a construction site and observe how the work stops as assorted workmen hang from the scaffolding to throw her verbal bouquets. The compliment has reached the level of an art form, and there is an annual contest with a prize for the best flowery accolade, according to category: classic, creative, erotic, comic, and poetic.

I was taught as a child to be discreet and to pretend to be virtuous. I say "pretend" because what you do but don't tell doesn't matter as long as no one finds out. In Chile we suffer from a particular form of hypocrisy. We act as if we're scandalized by any little peccadillo someone else commits at the same time that we are stacking up barbarous sins in private. We speak in euphemisms: "to nurse" is "to give the baby its milkie," and "torture" is referred to as "illegal pressure." We make a big show of being emancipated, but we are stoically silent about subjects considered taboo, not to be discussed publicly, from corruption (which we call "illicit enrichment") to film censorship, to mention only two. At one time *Fiddler on the Roof* was censored; now *The Last Temptation of Christ* is banned because of the opposition of the clergy and the fear that Catholic fundamentalists might set off a bomb in the theater. *Last Tango in Paris* made its appearance when Marlon Brando had become an obese old man and butter had gone out of style. The strongest taboo, especially for women, is still the taboo of sex.

The daughters of certain emancipated or intellectual families went to the university, but that was not true for me. My family thought of themselves as intellectuals but actually we were medieval barbarians. It was expected that my brothers would be professional men—if possible doctors or engineers, all other occupations were inferior—but I was to settle for a largely decorative job until motherhood occupied me completely. During those years, professional women came principally from the middle class, which is the strong backbone of the country. That has changed, I'm happy to say, and today the level of education for women is actually higher than for men. I wasn't a bad student, but since I already had a boyfriend it didn't occur to anyone that I might go to the university—not even to me. I finished high school at sixteen, so confused and immature that I had no idea what the next step might be, even though I always knew I would have to work because you can't be a feminist without financial independence. As my grandfather always said, the person who pays the bills rules the roost. I got a job as a secretary in one of the organizations of the United Nations, where I copied forestry statistics onto large pages of graph paper. In my free time I didn't embroider my trousseau, I read novels by Latin American authors and fought like a tiger with any male who crossed my path, beginning with my grandfather and my wonderful Tío Ramón. My rebellion against the patriarchal system was exacerbated when I went into the job market and found out for myself the disadvantages of being a woman.

And what about writing? I suppose that secretly I wanted to devote myself to literature, but I never dared put such a presumptuous goal into words, because that would have unleashed an avalanche of guffaws around me. No one had any interest in what I might have to say, much less write. I wasn't familiar with any important female authors, aside from two or three nineteenth-century English maiden ladies and our national female poet, Gabriela Mistral—but she was very mannish. Writers were mature men, solemn, remote, and usually dead. Personally I didn't know any, except for that uncle who went around the barrio playing the hurdy-gurdy, and who had published a book about his mystic experiences in India. Hundreds of copies of his thick novel were piled up in the cellar—bought, almost certainly, by my grandfather to get them out of circulation—fine material for the forts my brothers and I built when we were little. No, literature was definitely not a reasonable career path in a country like Chile where intellectual scorn for women was absolute. Through all-out war, we women have earned the respect of our troglodytes in certain areas, but the minute we're a bit careless, machismo raises its shaggy head again.

For a while I earned a living as a secretary, I married, and immediately became pregnant with my first child, Paula. Regardless of my feminist theories, I was a typical Chilean wife, selfless and servile as a geisha, the kind of woman who makes a baby of her husband, with premeditation and treachery. Enough to say, as proof, that I had three jobs, I ran the house, I looked after the children, and I ran like a marathoner the whole day to fight my way through the pile of responsibilities that had fallen on me, including a daily visit to my grandfather, but at night I waited for my husband with the olive for his martini between my teeth and the clothing he would wear the next morning carefully laid out. In any free moments, I shined his shoes and cut his hair and fingernails . . . just a run-of-the-mill Elvira.

Soon I was transferred to a different office, the department of information, where I was supposed to edit reports and act as press officer, either of which was much more entertaining than counting trees. I must admit that I didn't choose journalism, I was caught off guard; the profession simply sank its claws into me. It was love at first sight, a sudden passion that has determined a large part of my life. It happened during the early days of television in Chile, which consisted of two black-and-white channels originating from the universities. The only screen I'd ever seen was at the movies, but it was Stone Age TV, the most primitive stage, and though I hadn't taken regular courses at the university, I found myself launched upon a career. In those days, journalism was still a profession you learned on the job, and there was a certain tolerance for spontaneous practitioners like me. I should note here that in Chile women make up the majority among journalists, and are more prepared, visible, and courageous than their male colleagues; it is also true that they nearly always work under a man's orders. My grandfather was indignant when I told him what I was doing; he considered reporting an occupation for knaves; no one of sound mind would talk with the press, and no decent person would choose a calling in which the main order of work was talking about other people. However, I think he secretly watched my television programs because occasionally he let slip some revealing comment.

By the early sixties the rings of poor settlements around the capital city had grown in alarming fashion: cardboard walls, tin roofs, people in rags clearly visible along the road from the airport. Since this made a very bad impression on visitors, for a long time the solution was to put up walls to hide them. As one politician said, "Where there is poverty, hide it." There are marginal areas still today, despite the sustained effort of various governments to relocate the squatters in more decent barrios, but the situation is greatly improved. Back then, immigrants from the country and the most remote provinces came in massive numbers looking for work, and being unable to find decent housing they gravitated to these miserable hovels. Despite police harassment of the occupants, these shantytowns grew and became organized: once people took over a piece of land it was impossible to remove them or keep others from joining them. Shacks lined unpaved little streets that were a dust bowl in summer and a swamp in winter. Hundreds of barefoot children ran wild among the huts while parents went off every day to the city to look for a day's work that would "feed the pot," a vague term that could mean anything from earning a pitiable wage to buying a bone to make soup. Several times I visited these communities with a friend who is a priest, and afterward, when feminism and political unrest forced me out of my shell, I went often, trying to help. As a journalist I could make reports and tape interviews that helped me better understand our Chilean mentality.

Among the most acute problems, tied to the absence of hope, were alcoholism and domestic violence. Many times I saw women with battered faces. My sympathy fell on deaf ears because they always had an excuse for the aggressor: "He was drunk," "He got angry," "He was jealous," "If he hits me it's because he loves me," "Who knows what I did to provoke him?" I'm told that this situation hasn't changed much despite campaigns to prevent battering. In the lyrics of a popular tango, the man waits for his woman to fix his *mate* and then "knifes her thirty-five times." Police are now trained to burst into houses without waiting for the door to be opened normally, or before a corpse with thirty-five stab wounds is found hanging at the window, but there is still a long way to go. And we haven't even touched on the subject of child beatings! Every so often there is a story in the paper about some horrifying case of children tortured or beaten to death by their parents. According to the Inter-American Development Bank, Latin America is one of the most violent areas of the world, second only to Africa. Violence in the society begins at home; you can't eliminate crime in the streets unless you attack domestic aggression, since children who have been abused often become violent adults. Today there is a great deal of discussion on the subject, it is denounced in the press, and safe houses and education programs and police protection are available for victims, but in those days domestic crimes were taboo topics.

There was a strong class-consciousness in those squatter's settlements I visited, pride in belonging to the proletariat, which surprised me in a society as snobbish as Chile's. That's when I discovered that social climbing was a middle-class phenomenon, the poor never gave it a thought, they were too busy trying to survive. Over the years these communities acquired political savvy, they organized and became fertile territory for leftist parties. Ten years later, in 1970, they were decisive in electing Salvador Allende and for that reason had to suffer the greatest repression during the dictatorship.

I was very serious about journalism, even though colleagues from that time believe that I invented my reports. I didn't invent them, I merely exaggerated slightly. The experience left me with several obsessions: I find I am forever on the prowl for news and stories, always with a pencil and notepad in my handbag for jotting down anything that catches my eye. What I learned then helps now in my writing: working under pressure, conducting an interview, doing research, using the language efficiently. I never forget that a book is not an end in itself. Just like a newspaper or a magazine, a book is a means of communication, which is why I try to grab the reader by the throat and not let go to the end. I don't always succeed, of course; readers tend to be elusive. Who *is* my reader? Well, when the North Americans were in Panama and arrested General Noriega, who had fallen from grace, they found two books in his possession: the Bible and *The House of the Spirits.* You never know for whom you're writing. Every book is a message in a bottle tossed into the sea with the hope it will reach a different shore. I feel very grateful when someone finds it and reads it, particularly someone like Noriega.

In the meantime, Tío Ramón had been named the Chilean representative at the United Nations in Geneva. Letter exchange between my mother and me now took much less time than from Turkey, and occasionally it was possible to talk by telephone. When our daughter Paula was a year and a half old, my husband received a fellowship to study engineering in Belgium. On the map, Brussels looked very close to Geneva, and I didn't want to miss an opportunity to visit my parents. Ignoring the promise I'd made myself to put down roots and not go abroad for any reason, we packed our suitcases and set out for Europe. It was an excellent decision; among other reasons because I was able to study radio and television and renew my French, which I hadn't used since those days in Lebanon. During that year I discovered the Women's Lib movement, and realized that I wasn't the only witch in the world, there are many of us.

In Europe very few people had ever heard of Chile, but the country became fashionable four years later, with the election of Salvador Allende. It was in the news again in 1973 because of the military coup, then because of human rights violations, and eventually because of the arrest of the former dictator in London in 1998. Every time our country has made news, it has been for major political events, except for brief notes on the occasion of an earthquake. When someone in Europe asked my nationality in the sixties, I had to give long explanations and draw a map to demonstrate that Chile is at the southern tip of South America, not in the heart of Asia. It was often confused with China because of the somewhat similar name. The Belgians, used to the idea of colonies in Africa, were surprised that my husband spoke English and that I wasn't black. Once they asked me why I didn't wear traditional garb; they may have been thinking of Carmen Miranda's costumes in Hollywood movies: a multiruffled skirt and a basket of pineapples on her head. We traveled through Europe from Scandinavia to the south of Spain in a beat-up Volkswagen, sleeping in a tent and eating sausages, horse meat, and fried potatoes. It was a year of frenetic touring.

We returned to Chile in 1966 with our daughter Paula,

who at three spoke an academician's Spanish and had become an expert on cathedrals, and with Nicolás in my womb. In contrast with Europe, where long-haired hippies were a normal sight, student revolutions were brewing, and the sexual liberation was being celebrated, Chile was boring. Once again I felt like a foreigner, but I renewed my promise to grow roots and never leave.

As soon as Nicolás was born I went back to work, this time for a brand-new women's magazine called *Paula*. It was the only journal that promoted the feminist cause and featured subjects never aired until then, like divorce, contraception, domestic violence, adultery, abortion, drugs, and prostitution. Considering that in those days you couldn't say the word "chromosome" without blushing, we were suicidally audacious.

Chile is a hypocritical, prudish country bristling with scruples in respect to sex and sensuality, a nation of "old ladies," male and female. The double standard rules. Promiscuity is tolerated in men, but women must pretend that sex doesn't interest them, only love and romance, although in practice they must enjoy the same liberties as men—if not, who are the men dallying with? A female must never seem to be collaborating with the macho during the course of the seduction, she must be sly. It is supposed that if a girl is "difficult," the suitor's interest is held and she is respected; on the other hand, there are some very inelegant epithets for describing her reluctance. This is but a further manifestation of our hypocrisy, another of our rituals for maintaining appearances, because in truth there is as much adultery, as many teenage pregnancies, children born out of wedlock, and abortions, as in any other country. I have a woman friend who is a gynecologist and has specialized in looking after unmarried pregnant teenagers, and she assures me that unwanted pregnancies are much less common among university students. That happens more in low-income families, in which parents place more emphasis on educating and providing opportunities to their male children than to their daughters. These girls have no plans, they see a gray future, and they have limited education and little self-esteem; some become pregnant out of pure ignorance. They are surprised when they discover their condition because they have followed admonitions "not to go to bed with anyone" literally. What happened standing up, behind a door, surely didn't count.

More than thirty years have passed since *Paula* took a prudish Chilean society by storm, and no one can deny the effect of that hurricane. Each of the controversial articles in the magazine stirred my grandfather to the verge of cardiac arrest; we would argue at the top of our lungs, but the next day I would go back to see him and he would welcome me as if nothing had happened. In its beginnings, feminism, which today we take for granted, seemed extreme, and most Chilean women wondered why they needed it since they were already queens of their households and it was natural for men to be the bosses outside, the way God and Nature had intended. It was hard work to convince them that they weren't queens anywhere. There were not many visible feminists; at the most, half a dozen. I try not to remember what aggravation we had to put up with! I realized that to wait to be respected for being a feminist was like expecting the bull not to charge because you're a vegetarian. I also went back to television, this time with a comedy show, and while doing that acquired a certain visibility, as happens to anyone who appears regularly on the screen. Soon every door was open to me, people greeted me in the street, and for the first time in my life I felt I belonged.

DISCREET CHARM OF THE BOURGEOISIE

I often ask myself what exactly nostalgia *is.* In my case, it's not so much wanting to live in Chile as it is the desire to recapture the certainty I feel there. That's my home ground. Each country has its customs, its manias, its complexes. I know the idiosyncrasies of mine like the palm of my hand; nothing surprises me, I can anticipate others' reactions, I understand what gestures mean, silences, formulas of courtesy, ambiguous responses. Only there do I feel comfortable socially—despite the fact I rarely behave as I'm expected to—because there I know how to behave and my good manners rarely fail me.

When I was a recently divorced forty-five, I immigrated to the United States, obeying the call of my impulsive heart. The first thing that surprised me was the infallible optimism of North Americans, so different from people in the southern tip of South America, who always expect the worst to happen. Which it does, of course. The U.S. Constitution guarantees the right to the pursuit of happiness, which anywhere else would be an embarrassing presumption. North Americans also believe they have the eternal right to be entertained, and if any of their rights are denied, they feel frustrated. The rest of the world, in contrast, expects that on the whole, life is hard, and boring, so they celebrate sparks of joy and diversion, however modest, when they occur.

In Chile it is bad manners to acknowledge that you're overly satisfied, because that can irritate the less fortunate, which is why for us the correct answer to the question "How are you?" is "So-so." That is an opening for sympathizing with the other speaker's situation. For example, if one person says he's just been diagnosed with a fatal illness, it would be very bad taste to rub in his face how well everything's going for you, wouldn't it? But if the other person has just married an heiress, you're free to confess your own happiness without fear of wounding anyone's feelings. That is the sense of the "So-so" that can sometimes confuse visiting foreigners: it gives us time to feel out the ground and avoid a faux pas. Sociologists say that forty percent of Chileans suffer from depression, especially women, who have to put up with the men. You must remember, too, that our country goes through major disasters, and that there are many poor, so it seems rude to mention one's own good fortune. I had a relative who twice won the jackpot in the lottery, but he always said "So-so," in order not to offend. As an aside, it's rather interesting to learn how his good fortune came about. He was a very strong Catholic and as such never wanted to hear talk of contraceptives. After his seventh child was born, desperate, he went to the church, knelt before the altar, and had a heart-to-heart talk with his Creator. "Lord, since you sent me seven children, it would be a kindness if You helped me feed them," he argued, and immediately took a long, carefully prepared list of expenses from his pocket. God listened patiently to the arguments of his loyal servant and almost immediately revealed the winning lottery number in a dream. Those millions lasted for several years, but inflation, which was endemic in Chile during that time, reduced his capital at the same rate he enlarged his family. When the last of his children was born, number eleven, he returned to church to argue his case, and again God came to his aid by sending another revelation in a dream. The third time it was no deal.

In my family, happiness was irrelevant. My grandparents, like the great majority of Chileans, would have stood with their mouths agape if they'd known that there are people who spend good money on therapy to overcome their unhappiness. For them, life was just difficult, any other view was foolishness. You found satisfaction in doing the right thing, in family, honor, the spirit of service, study, and your own fortitude. Joy was in our lives in many ways, and I suppose that love was not the least important, but we didn't talk about it, we would have died of shame before saying the word. Emotions flowed silently. In contrast to most Chileans, in our family we didn't touch much and babies were never coddled. The modern custom of extolling a child's every move as if it were witty and charming was not in vogue, nor was there anxiety about bringing up offspring who were free of traumas. Just as well, because if I'd been brought up protected and happy, what the devil would I write about now? With this in mind, I've tried to make my grandchildren's childhood as difficult as possible so they will grow up to be creative adults. Their parents are not at all appreciative of my efforts.

Physical appearance was ignored in my family; my mother swears that she didn't know she was pretty until she was forty, because looks were never mentioned. In that, we could claim originality because in Chile appearances are fundamental. In our clan it was also bad taste to talk about religion and, most of all, money. On the other hand, illness was a constant topic of conversation, it is Chile's most chewed-over topic. We specialize in exchanging remedies and medical advice; everyone loves to prescribe a cure. We distrust doctors because it's obvious that good health does not promote good business, and we go to them only when everything else has failed, after we've tried all the remedies recommended by our friends and acquaintances. Let's say you faint at the door of a supermarket. In any other country they call an ambulance, but not Chile, where several volunteers will pick you up, haul you behind the checkout counter, pour cold water on your face and whiskey down your gullet to bring you to; then they will force you to swallow pills some lady takes from her purse because "my friend has these attacks and this is a fantastic remedy."

There will be a chorus of experts who will diagnose your condition in clinical terms because every citizen with an ounce of sense knows a lot about medicine. One of the experts, for example, will say that you have an obturation of a valve in your brain, but another may suspect a complex torsion of the lungs, and a third that you have ruptured your pancreas. Within a few minutes there will be a hue and cry all around you, and someone will arrive who's run to the pharmacy to buy penicillin to inject you with—just in case. Come to think of it, if you're a foreigner, my advice is not to faint in a Chilean supermarket; it can be a deadly experience.

To illustrate how free we are about prescribing, once during a southern cruise to our beautiful San Rafael lagoon in the cold fjords of the south, we were given sleeping pills with dessert. At dinner the captain notified the passengers that we were about to sail through particularly rough waters, and then his wife went from table to table handing out pills, the name of which no one dared ask. We took them obediently and twenty minutes later all the passengers were out like a light, suggesting the story of Sleeping Beauty. My husband said that in the United States the captain and his wife would have been arrested for anaesthetizing the passengers. In Chile we were very grateful.

In times gone by, the minute two or more people got together, the obligatory subject was politics; if there were two Chileans in a room, you could be sure of finding three political parties. I understand that in one period we had more than a dozen socialist mini-parties; even the right, monolithic in the rest of the world, was split. However, politics no longer brings out our passions; we talk about it only to be able to complain about the government, one of our favorite activities. We no longer vote religiously, as in the days when dying citizens were carried on stretchers to fulfill their civic duty. Nor do we, as we once did, have instances of women giving birth in the voting booth. The young don't register to vote, some 84.3 percent of the people believe that political parties do not represent their interests, and a greater number say they are content not to participate in any way in the conduct of the nation. This is a phenomenon of the Western world, it appears. Young people have no interest in fossilized political schemes dragged over from the nineteenth century. They are preoccupied with living well and prolonging their teenage years as long as possible— let's say, till about forty or fifty. To be fair, there is also a small percentage who are militants in respect to ecology, science, and technology, and I have heard about some who do social service through the churches.

The subjects that have replaced politics among Chileans are money, which there is never enough of, and soccer, which is a kind of consolation. The lowliest illiterate knows the names of all the players throughout our history, and has his own opinion of each. This sport is so important that souls from purgatory wander the streets freely when there's a match because the entire population is in a catatonic state in front of television sets. Soccer is one of the few human activities that proves the relativity of time: the goalie can float in the air for half a minute, the same scene can be repeated several times in slow motion, or backward and, thanks to the time change between continents, a game between Hungary and Germany can be seen in Santiago before it's played.

In our house, as in the rest of the country, dialogue was unknown; our get-togethers consisted of a series of simultaneous monologues during which no one listened to anyone: pure confusion and static, like a short-wave radio transmission. It didn't matter, because neither was there interest in learning what others thought, only in repeating one's own side of things. When my grandfather grew old, he refused to wear a hearing aid because he thought that the only thing good about his years was not having to listen to the foolish things people said. As General Mendoza expressed so eloquently in 1983: "We are abusing dialogic expression. There are cases in which dialogue is unnecessary. A monologue is more necessary because a dialogue is a simple conversation between two people." This philosopher added later that "The country lives in organized disorder." My family would have been in total agreement.

We Chileans have a tendency to speak in falsetto. Mary Graham, an Englishwoman who visited the country in 1822, commented in a book titled *Diary of My Residence in Chile* that people were charming, but that they spoke in a disagreeable tone of voice, especially the women. We swallow half our words, we aspirate the *s* and change vowels, so that the word *señor* sounds like *inyol*. There are at least three official languages: the educated speech used in communication media, in official matters, and by some members of the upper class when not among friends; the colloquial language used by ordinary people; and the indecipherable and always changing speech of young people. The visiting foreigner should not despair, because even if he doesn't understand a word, he'll see that people are dying to be of help. We also speak very low and sigh a lot. When I lived in Venezuela, where men and women are very sure of themselves and of the ground beneath their feet, it was easy to distinguish my compatriots by the way they walked—like spies in disguise—and by their unvarying tone of apology. I used to go every morning to a Portuguese bakery to have my first cup of coffee, where there was always a mob of customers fighting to get to the counter. The Venezuelans would shout from the door, "Hey, a coffee over here!" and before you knew it they had a paper cup of *café con leche,* passed to them hand to hand. We Chileans—and in that period there were a lot of us because Venezuela was one of the Latin American countries that accepted refugees and immigrants—would hold up a trembling index finger and in the thread of a voice plead, "I'm sorry, may I please have a little cup of coffee, señor?" We could stand there the entire morning, waiting in vain. The Venezuelans joked about our much too precious manners, and in turn we Chileans were shocked by how forward they were. Those of us who lived in that country for several years changed, though, and among other things we learned to shout when we ordered our coffee.

. . .

Having read these observations about the character and customs of Chileans, you can understand my mother's doubts: there's no reason I should have turned out the way I did. I have none of the sense of decorum, the modesty, or the pessimism of my relatives, and none of their fear of what people will say, of extravagance, or of God. I don't speak or write apologetically, instead I'm rather grandiloquent, and I like attracting attention. That is, I simply am as I am today, after a lot of living. In my childhood I was a strange little insect; in adolescence, a shy mouse—for many years my nickname was *Laucha,* which was what we called our ordinary household mice—and in my youthful years I was everything from a rabid feminist to a flower-crowned hippie. My worst flaw is that I tell secrets, my own and everybody else's. In short, a disaster. If I lived in Chile no one would speak to me. But one thing I am is hospitable. At least they managed to hammer that virtue into me when I was a child. Knock at my door at any hour of the day or night and even if I've just broken my femur I will crawl to open the door and offer you your first cup of tea of the day. In everything else, I am the antithesis of the lady my parents, with great sacrifice, tried to make of me. It isn't their fault, they simply had very little to work with, and besides, I was bent by destiny.

If I had stayed in Chile, as I always wanted, married to one of my second cousins (in the improbable case that one of them proposed to me), maybe today I would carry my ancestors' blood in my veins with dignity, and perhaps my father's coat of arms bearing the flea-bitten dogs would be hanging in a place of honor in my home. I should add that however rebellious I may have been in my life, I have not lost the manners that were drummed into me day and night, as they should be for anyone who is to be a "decent" person. Which was fundamental in my family. That word encompassed much more than I could possibly explain in these pages, but I can say with absolute confidence that courtesy and good manners were a large part of what was defined as decency.

Well, I've gone way off on a tangent, and I need to pick up the main thread of this account, if there is any thread in all this meandering. But that's how nostalgia is: a slow dance in a large circle. Memories don't organize themselves chronologically, they're like smoke, changing, ephemeral, and if they're not written down they fade into oblivion. I've tried to arrange my thoughts according to themes or periods of my life, but it's seemed artificial to me because memory twists in and out, like an endless Moebius strip.

A BREATH OF HISTORY

A nd since we're talking about nostalgia, I beg you to have a little patience with what follows because I can't separate the subject of Chile from my own life. My past is composed of passions, surprises, successes, and losses: it isn't easy to relate in two or three sentences. I suppose there are moments in all human lives in which our fate is changed or twisted and forced to follow a different course. That has happened several times in mine, but maybe one of the most defining was the military coup in 1973. Were it not for that event, it's clear that I would never have left Chile, that I wouldn't be a writer, and that I wouldn't be married to an American and living in California. Nor would I have lived with nostalgia for so long, or be writing these particular pages. All of which leads inevitably to the theme of politics. To understand how the military coup could have come about, I must briefly refer to our political history, from its beginnings to the time of General Augusto Pinochet, who today is a senile old man living under house arrest, but nonetheless a man whose importance it is impossible to ignore. More than one historian considers Pinochet to be the most singular political figure of the twentieth century, though that is not necessarily a favorable judgment.

In Chile the political pendulum has swung from one extreme to another; we have tested every system of government that exists, and we have suffered the consequences. It isn't strange, therefore, that we have more essayists and historians per square foot than any nation in the world. We study ourselves incessantly; we have the vice of analyzing our reality as if it were a permanent problem requiring urgent solutions. The brains who burn the midnight oil in this pursuit are a bunch of tedious eggheads who say things no one understands a single word of; as a result, no one pays much attention to them. Among us Chileans, pessimism is considered good form; it is assumed that only idiots go around happy. We are a developing nation, the most stable, secure, and prosperous in Latin America and one of the most organized; no one surpasses us in character, but it is very annoying to us when someone decrees that "the country is in fine shape." Anyone who dares say that must be considered an ignoramus who never reads the newspapers.

Ever since its independence in 1810, Chile has been run by the social class that has the economic power. Formerly that was landowners; today it is entrepreneurs, industrialists, and bankers. Formerly the powerful belonged to a small oligarchy that had descended from Europeans and was composed of a handful of families; today the ruling class is broader, numbering several thousand of the kinds of persons who know how to get things done. During the first hundred years of the republic, the presidents and politicians were all from the upper class, though later the middle class also had a hand in governing. Few, nevertheless, came from the working class. Presidents with a social conscience were men moved by inequality, injustice, and poverty, even though they had not experienced those afflictions personally. Today the president and the majority of politicians, with the exception of several rightists, are not members of the economic group that has true control of the country. At this moment we have a paradoxical situation: a Socialist president and a rightist economy and policies.

Until 1920 the country was ruled by a conservative oligarchy with a feudal mentality. One exception was the liberal president José Manuel Balmaceda (1891) who perceived the needs of the people and who tried to bring about reforms that threatened to damage the interests of the landowners, though he himself came from a very wealthy and powerful family, owners of an enormous latifundio. The conservative parliament's fierce opposition provoked a social and political crisis. The navy rebelled and allied itself with the parliament, and a cruel civil war was unleashed that ended with the triumph of the parliament and with Balmaceda's suicide. Nevertheless, the seeds of social ideas had been planted, and the following years saw the birth of the radical and communist parties.

In 1920 a political leader was elected who for the first time preached social justice: Arturo Alessandri Palma, nicknamed The Lion. He came from a middle-class family of second-generation Italian immigrants. Although his family wasn't wealthy, his European heritage, his culture, and his education easily qualified him for a place in the ruling class. He promulgated social legislation, and during the term of his government workers organized and gained access to political parties. Alessandri suggested modifying the Constitution to establish a true democracy, but the conservative forces of the opposition impeded him from accomplishing that, even though the majority of Chileans, especially the entire middle class, supported him. Parliament (again the parliament!) made it difficult for him to govern; it forced him to resign his position and exiled him to Europe. A succession of military juntas attempted to govern, but the country seemed to lose direction and the popular outcry forced the return of The

Lion, who ended his term by seeing a new constitution put into effect.

The armed forces, which felt they had been eased out of power and believed that the country owed them a great deal given their victories during the wars of the preceding century, forcefully installed General Carlos Ibáñez de Campo in the office of president. Ibáñez quickly employed dictatorial measures—which have continued to be anathema to Chileans up to the present moment—and that produced a civil opposition so formidable that it paralyzed the country and the general had to resign. Then came a period that we can classify as a sane democracy. Alliances were formed among parties, and in 1938 the left came to power under President Pedro Aguirre Cerda, a member of the Frente Popular, or Popular Front, in which communist and radical parties participated. After Pedro Aguirre Cerda, the deposed Ibáñez joined forces with the left, and three successive radical presidents followed. (Even though I was just a girl at the time, I remember that when Ibáñez was elected to govern for the second time, my family went into mourning. In my hideaway beneath the grand piano I heard the apocalyptic prognostications of my grandfather and uncles. I spent sleepless nights, convinced that enemy hordes were coming to burn our house to the ground. No such thing happened, the general had learned his lesson and acted within the Constitution.) For twenty years we had center-left governments, until 1958, when the right triumphed with Jorge Alessandri, son of The Lion and completely different from his father. The Lion was a populist with advanced ideas for his time; his son was a conservative, and projected an old-maidish image.

While revolutions were erupting in most Latin American countries and caudillos were taking over governments at gunpoint, an exemplary democracy was being consolidated in Chile. The first half of the twentieth century witnessed the crystallization of significant social advances. Free, public, compulsory education, public health for all, and one of the most advanced social security systems on the continent favored the strengthening of a vast educated and politicized middle class, as well as a proletariat with class awareness. Unions were formed, along with centers for workers, employees, and students. Women gained the vote, and electoral processes were perfected. (An election in Chile is as civilized as tea time in London's Savoy Hotel. Citizens line up in queues to vote, without ever producing the least altercation, even if political tempers are boiling. Men and women vote at different sites, guarded by soldiers to avoid disturbances or bribery. No alcohol is sold the previous day, and businesses and offices remain closed. No one works on Election Day.)

Concern for social justice also reached into the Catholic Church, which has great influence in Chile, and which on the basis of new encyclicals made great efforts to support the changes being effected in the country. In the meantime two large blocks of influence were being affirmed in the outside world: capitalism and socialism. To confront Marxism, the Christian Democrats were born in Europe, a center-right party with a humanist and community-oriented message. In Chile, where it promised "revolution with freedom," that party destroyed the opposition in the election of 1964, defeating both the conservative right and parties on the left. The overwhelming triumph of Eduardo Frei Montalva, and a Christian Democrat majority in the parliament, marked a milestone. The country had changed. It was assumed that the right had faded into history, that the left would never have its chance, and that Christian Democrats would govern till the end of time, but that did not happen, and after only a few years the party lost popular support. The right had not been shattered, as had been predicted, and the left came back from defeat and reorganized. Power was divided into three parts: right, center, and left.

At the end of Frei Montalva's term, the country was in an uproar. There was a suffocating atmosphere of revenge on the part of the right, which felt its wealth had been expropriated and which feared it would definitively lose the power it had always boasted of, and of resentment on the part of the lower classes, which had never felt represented by the Christian Democrats. Each of the three segments of power presented a candidate: Jorge Alessandri for the right, Radomiro Tomic for the Christian Democrats, and Salvador Allende for the left.

The parties of the left joined together in a coalition called the Unidad Popular, which included the Communist Party. The United States was alarmed despite the results of polls giving the victory to the right, and it designated several million dollars for defeating Allende. The political forces were so divided that Allende, with his theme of "the Chilean route to socialism," won by a narrow margin, with 38 percent of the vote. Since he did not obtain an absolute majority, the election would have to be ratified by the Congress, which traditionally had given the nod to the candidate with the most votes. Allende was the first Marxist to win the presidency of a country through a democratic vote. The eyes of the world turned toward Chile.

Salvador Allende Gossens was a charismatic physician who had been minister of health in his youth, senator for many years, and also the eternal presidential candidate of the left. He himself told the joke that on his death his epitaph would read "Here lies the next president of Chile." He was courageous, loyal to friends and collaborators, magnanimous to his adversaries. He was considered vain because of the way he dressed, and because of his taste for the good life and beautiful women, but he was deeply serious in regard to his political convictions. In that, no one can accuse him of frivolity. His enemies preferred not to confront him personally, because he had the reputation of being able to manipulate any situation to his benefit. He endeavored to institute profound economic reforms within the frame of the Constitution, to expand the agrarian reform initiated by the previous government, and to nationalize the private enterprises, banks, and copper mines that were in the hands of North American companies. He proposed a socialist system that respected civil rights, an experiment that no one had attempted before.

The Cuban revolution had by that time survived ten

years, despite the efforts of the United States to destroy it, and there were leftist guerrilla movements in many Latin American countries. The undisputed hero of young people was Che Guevara, who had been assassinated in Bolivia, and whose face—picture a saint wearing a beret and smoking a cigar—had become a symbol of the struggle for justice. Those were the days of the Cold War, when an irrational paranoia divided the world into two ideologies and determined the foreign policies of the Soviet Union and United States for several decades. Chile was one of the pawns sacrificed in that conflict of titans. The administration of Richard Nixon decided to intervene directly in the Chilean process. Henry Kissinger, who was responsible for foreign policy, and who admitted he knew nothing about Latin America, which he considered the "backyard" of the United States, said that "there was no reason to watch as a country became communist through the irresponsibility of its own people, and do nothing about it." (This joke circulates around Latin America: Do you know why there are no military coups in the United States? Because there's no North American embassy.) To Kissinger, Salvador Allende's democratic path toward Socialism seemed more dangerous than an armed revolution because of the danger of infecting the rest of the continent like an epidemic.

The CIA orchestrated a plan to prevent Allende from assuming the presidency. First it tried to bribe members of Congress not to designate Allende and to call for a second vote in which there would be only two candidates: Allende and a Christian Democrat supported by the right. Since the bribes didn't work, the CIA planned to kidnap the commander in chief of the armed forces, General René Schneider; although the plot would be carried out by a neo-Fascist group, it would appear to be the work of a leftist commando unit. The idea was that this action would provoke chaos and a military intervention. The general was shot to death in the skirmish, but the plan had the opposite of the desired effect: a wave of horror washed across the country and the Congress unanimously awarded Salvador Allende the presidential sash. From that moment on, the right and the CIA plotted together to oust the government of the Unidad Popular, even at the cost of destroying the economy and Chile's long democratic tradition. Then the CIA activated an alternate plan: a so-called destabilization, which consisted of cutting off international credit and initiating a campaign of sabotage to incite economic ruin and social violence. Simultaneously, their siren song was directed at the military, which in the end held the strongest card in the game.

The right, which controls the press in Chile, organized a campaign of terror that included posters with Soviet soldiers ripping babies from their mothers' arms to be taken to the gulags. On Election Day, when it was apparent that Allende had triumphed, people came out in force to celebrate: never had there been such a huge popular demonstration. The rightists had ended up believing their own propaganda, and barricaded themselves in their houses, convinced that inflamed *rotos* were coming to commit unimaginable atrocities. The euphoria of the common people was extraordinary—signs, banners, embraces—but there were no excesses and at dawn the celebrators retired to their homes, hoarse from singing. The next day there were long lines in front of the banks and travel agencies in the upper-class barrio: many people withdrew their money and bought tickets to flee abroad, convinced that the country was going down the same road as Cuba.

Fidel Castro arrived to show his support for the Unidad Popular and that exacerbated the opposition's panic, especially when it saw the reception given the controversial Comandante. Organized by labor and professional unions, schools, political parties, and others, people lined up along the highway from the airport to the center of Santiago; there were banners and standards and marching bands, in addition to a huge anonymous crowd that went to watch the spectacle out of curiosity, with the same enthusiasm that years later they would lavish on the pope. The visit of the bearded Comandante lasted too long: twenty-eight endless days, during which he traveled the country from north to south, accompanied by Allende. I believe we all gave a sigh of relief when he left, but it can't be denied that his retinue left the air filled with music and laughter. Cubans are enchanting; twenty years later I came to know some exiled Cubans in Miami, and found that they are as pleasant as the islanders. We Chileans, always so serious and solemn, were shaken by the whole experience: we didn't know that life and revolution could be lived with such joy.

The Unidad Popular was popular, but it wasn't united. The parties in the coalition fought like cats and dogs for every morsel of power, and Allende had to confront not only the opposition on the right but also critics in his own ranks who demanded swifter and more radical action. Workers took over factories and farms, weary of waiting for the nationalization of private enterprise and the expansion of agrarian reform. Sabotage by the right, North American intervention, and errors on the part of Allende's government provoked a grave economic, political, and social crisis. Inflation rose officially to 360 percent a year, although the opposition claimed it was more than 1000 percent, which meant that a housewife woke up every morning not knowing how much bread would cost that day. The government fixed the prices of basic products, and many industries and agribusinesses failed. The shortages were so severe that people spent hours waiting to buy a scrawny chicken or a cup of cooking oil, but those who could pay bought anything they wanted on the black market. With their modest way of talking and behaving, Chileans referred to a queue as *la colita*, "the mini-line," even when it was three blocks long, and sometimes stood in them without knowing what was being sold, just out of habit. Soon there was a psychosis of shortages, and as soon as three or more people were together, they automatically started a queue. That was how I once bought cigarettes, though I've never smoked, and another time ended up with eleven tins of colorless shoe wax and a gallon of soy extract I can't imagine a use for. There were professional line-standers who got tips for holding a place; I understand that my own children rounded out their allowance that way.

Despite the problems and the climate of permanent confrontation, ordinary people were excited because for the first time they felt they had some control over their destinies. A true renaissance took place in the arts, folklore, and popular and student movements. Masses of volunteers went out to eradicate illiteracy in every corner of Chile; books were published at the price of a newspaper, so there was a library in every house. For their part, the economic right, the upper class, and a sector of the middle class—particularly housewives, who suffered the problems of shortages and loss of order—detested Allende and feared that he would be perpetuated in government as Fidel Castro had been in Cuba.

Salvador Allende was my father's cousin and the only person in the Allende family who kept in contact with my mother after my father left. He was a good friend of my stepfather, so I had several opportunities to be with him during his presidency. Although I didn't take part in his government, those three years of the Unidad Popular were surely the most interesting in my life. I have never felt so alive, nor have I ever again participated so closely in a community or in the life of a nation.

From a contemporary perspective, we can agree that Marxism as an economic project is dead, but I think that some of Salvador Allende's principles are still attractive, particularly his search for justice and equality. He was trying to establish a system that would give the same opportunities to everyone and create "the new man," who would act for the common good, not personal gain. We believed that it was possible to change people through indoctrination; we refused to see that in other countries, even where they had tried to impose a system with an iron hand, the results were very doubtful. The sudden breakdown of the Soviet system was still in the future. The premise that human nature is susceptible to such a radical change now seems ingenuous, but then, for many of us, it was the ultimate goal. This ideal blazed like a bonfire in Chile. Typical Chilean characteristics, such as sobriety, a horror of ostentation, of standing out over others or attracting attention, generosity, a tendency to compromise rather than confront, a legalistic mentality, respect for authority, resignation to bureaucracy, enthusiasm for political argument, and many others, found their perfect home in the Unidad Popular. Even fashion was affected. During those three years, models in women's magazines were dressed in rough workman's textiles and clunky proletarian shoes, and bleached flour-sacking was used to make blouses. I was responsible for the home decorating section of the magazine where I worked, and my challenge was to produce photographs of attractive and pleasing décors achieved with minimal cost: lamps made from large tins, rugs woven of hemp, pine furniture darkened with stain and burned with a blowtorch to look antique. We called it the "monastery mode," and the idea was that anyone could knock out these pieces at home with four boards and a saw. It was the golden age for the DFL2 Act, which allowed buyers to acquire houses of a maximum of one hundred and forty square meters at low cost and with tax breaks. Most houses and apartments were the size of a two-car garage; ours had ninety square meters and we thought it was a palace. My mother, who was in charge of the cooking section of *Paula,* had to invent inexpensive recipes that didn't call for scarce ingredients; however, bearing in mind that everything was scarce, her creativity was rather restricted. One Peruvian artist who arrived for a visit during that period asked, amazed, why Chilean women dressed like lepers, lived in doghouses, and ate like fakirs.

Despite the many problems the population faced during that time, from those multiple shortages to political violence, three years later in the parliamentary elections of March 1973, the Unidad Popular increased its margin of votes. Efforts to derail the government through sabotage and propaganda had not had the hoped-for results. That was when the opposition moved into the last phase of the conspiracy and incited a military coup. We Chileans had no idea what that entailed, because we had a long and solid democratic tradition and we were proud of being different from other countries of the continent, which we scornfully referred to as "banana republics," where every other day some caudillo took over the government by force. No, that would never happen to us, we proclaimed, because in Chile even the soldiers believed in democracy, no one would dare violate our Constitution. That was pure ignorance; if we had looked back over our history, we would have been better acquainted with the military mentality.

When I did the research for my novel *Portrait in Sepia,* published in English in 2001, I learned that in the nineteenth century our armed forces waged several wars, giving evidence of as much cruelty as courage. One of the most famous moments of our history was the capture of the Arica promontory in June 1880, during the War of the Pacific against Peru and Bolivia. That stronghold was an impregnable cliff with a two-hundred-meter drop straight to the sea; there large numbers of Peruvian troops equipped with heavy artillery were defended by three kilometers of sand bags and a surrounding mine field. The Chilean soldiers launched their attack with curved knives between their teeth and bayonets bared. Many fell beneath enemy fire or were blown to bits by exploding mines, but nothing stopped their remaining comrades, who climbed up to the fortifications and over them, thirsty for blood. They gutted Peruvians with knife and bayonet and took the headland, an incredible feat that lasted only fifty-five minutes. Then they killed the conquered, finished off the wounded, and sacked the city of Arica. One of the Peruvian commanders jumped into the sea rather than fall into the hands of the Chileans. The figure of the dashing officer and his black steed with its legendary gold horseshoes leaping from the cliff is part of the legend of that ferocious episode. The war was decided later with Chile's triumph at the battle of Lima—which Peruvians remember as a massacre, even though Chilean history books claim that our troops occupied the city in an orderly fashion.

The victors write history in their own way. Every country presents its soldiers in the most favorable light, hides their mistakes and downplays their atrocities, and after the battle is won everyone is a hero. Since we grew up with the idea that the Chilean armed forces were composed of obedient soldiers under the command of irreproachable officers, we were in for a tremendous surprise that Tuesday, September 11, 1973, when we saw them in action. Their savagery was so extreme that it's believed they were drugged, just as the men who took the Arica promontory were intoxicated with *chupilca del diablo,* an explosive mix of liquor and gunpowder. In 1973, the army surrounded the Palacio de la Moneda, the seat of government and symbol of our democracy, with tanks, and then its planes bombed it from the air. Allende died inside the palace; the official version is that he committed suicide. There were hundreds of dead and so many thousands of prisoners that the sports stadiums and even some schools were turned into jails, torture centers, and concentration camps. Using the pretext of liberating the country from a hypothetical Communist dictatorship that might occur in the future, democracy was replaced by a regimen of terror that was to last sixteen years, and leave its consequences for a quarter of a century.

I remember fear as a permanent metallic taste in my mouth.

GUNPOWDER AND BLOOD

To give an idea of what the military coup was like, you have to imagine how a citizen of the United States or Great Britain would feel if the army rolled up in full battle gear to attack the White House or Buckingham Palace, and in the process caused the deaths of thousands of citizens, among them the president of the United States or the queen and the prime minister of Great Britain, then indefinitely suspended Congress or Parliament, disbanded the Supreme Court, abrogated individual liberties and political parties, declared absolute censorship of the media, and finally, over time, strove mercilessly to extinguish every dissident voice. Now imagine that these same military men, possessed with messianic fanaticism, installed themselves in power for years, prepared to root out every last ideological adversary. That is what happened in Chile.

The socialist adventure ended tragically. The military junta, presided over by General Augusto Pinochet, applied the doctrine of "savage capitalism" as the neoliberal experiment has been called, but refused to acknowledge that to function smoothly it requires a labor force free to exercise its rights. Brutal repression was used to destroy the last seed of leftist thought and implant a heartless capitalism. Chile was not an isolated case, the long night of dictatorships darkened the continent for more than a decade. In 1975, half of Latin America's citizens lived under some kind of repressive government, most of which were backed by the United States, which has a shameful record of overthrowing legally elected governments and of supporting tyrannies that would never be tolerated in its own territory: Papa Doc in Haiti, Trujillo in the Dominican Republic, Somoza in Nicaragua, and many others.

I realize as I write these lines that my view is subjective. I should report events dispassionately, but that would be to betray my convictions and sentiments. This book is not intended to be a political or historical chronicle, only a series of recollections, which always are selective and tinted by one's own experience and ideology.

The first stage of my life ended that September 11, 1973. I won't expand on that here since I have already recounted it in the final chapters of my first novel and in my memoir *Paula*. The Allende family—that is, those who didn't die— were taken prisoner, went into hiding, or left the country. My brothers, who were out of the country, did not return. My parents, who were in the embassy in Argentina, remained in Buenos Aires for a while, until they received death threats and had to escape. Most of my mother's family, on the other hand, were bitterly opposed to the Unidad Popular, and many of them celebrated the military coup with champagne. My grandfather detested socialism and eagerly awaited the end of Allende's government, but he never wanted it to be at the cost of democracy. He was horrified to see the government in the hands of the military, whom he despised, and he ordered me not to get involved. It was impossible, however, for me to stay on the edges of what was happening. This fine old man spent months observing me and asking tricky questions; I think he suspected that his granddaughter would vanish at any moment. How much did he know about what was happening around him? He lived an isolated life, he almost never went out of the house, and his contact with reality came through the press, which suppressed the truth and overtly lied. I may have been the one person who gave him the other side of the picture. At first I tried to keep him informed because in my role as a journalist I had access to the underground network that replaced serious sources of information during that period, but eventually I stopped bringing him bad news because I didn't want to frighten or depress him. Friends and acquaintances began to disappear; some returned after weeks of absence, with the eyes of madmen and signs of torture. Many sought refuge in other countries. In the beginning, Mexico, Germany, France, Canada, Spain, and other countries took them in, but after a while they had to call a halt because thousands of other Latin American exiles were being added to the waves of Chileans.

In Chile, where friendship and family are very important, something happened that can be explained only by the effect fear has on the soul of a society. Betrayal and denunciation snuffed out many lives; all it took was an anonymous voice over the telephone for the badly named intelligence services to sink their claws into the accused, and in many cases nothing was ever heard of that person again. People were divided between those who backed the military government and those who opposed it; hatred, distrust, and fear poisoned relationships. Democracy was restored more than a decade ago, but that division can still be felt, even in the heart of many families. Chileans learned not to speak out, not to hear, and not to see, because as long as they were not aware of events, they didn't feel they were accomplices. I know people for whom Allende's government represented the most unstable and dangerous state of affairs that could befall a country. For them, individuals who pride themselves on leading their lives in accord with strict Christian principles, the need to destroy that rule was so imperative that they didn't question the methods. Not even when a desperate father, Sebastián Acevedo, poured gasoline over himself and set himself on fire in Concepción Plaza, immolating himself like a Buddhist monk to protest the torture of his children. Ways were found to ignore—or pretend to ignore—violations of human rights for many years, and, to my surprise, I still find some who deny those crimes occurred, despite all the evidence. I can understand them because they are as adamant regarding their beliefs as I am mine. Their opinion of Allende's government is nearly identical to mine of the dictatorship of Pinochet, with the difference that in my view the end does not justify the means. Crimes perpetrated in shadows during those years have, inevitably, been coming to light. Airing the truth is the beginning of reconciliation, although the wounds will take a long time to heal because those responsible for the repression have not admitted their guilt and are not disposed to ask forgiveness. The acts of the military regime will go unpunished, but they can no longer be hidden or ignored. Many, especially young people who grew up without political dialogue or without a critical spirit, believe that there's been enough digging through the past, that we must look to the future, but victims and their families cannot forget. It's possible that we will have to wait until the last witness to those times dies before we can close that chapter of our history.

The military who took power were not models of culture. Seen from the perspective of the passage of years, the things they said are laughable; at the moment they were spoken they were quite terrifying. Exaltation of the nation, of "Western Christian values," and of militarism reached ridiculous levels. The country was run like a barracks. For years I had written a humor column in a magazine and anchored a lighthearted program on television, but nothing like that could work in that atmosphere because in truth there was nothing to laugh about—except those who were governing, which would have cost you your life. The one sliver of humor may have been *Tuesdays with Merino.* One of the high-ranking members of the junta, Admiral José Toribio Merino, met weekly with the press to offer opinions on assorted topics. The journalists eagerly awaited these pearls of mental clarity and acumen. For example, in regard to the 1980 change in the Constitution that was intended to legalize the military's assault on power, he stated with the greatest seriousness that "The first transcendence I see in it is that it is transcendent." And then he immediately explained, so everyone would understand: "There were two criteria in the creation of this Constitution: the political criterion, let us say, Platonic-Aristotelian in the classic Greek sense, and the second, the absolutely military criterion, which comes from Descartes, which we shall call Cartesian. In Cartesianism, the Constitution meets all that, the kind of definitions that are extraordinarily positive, which look for truth without alternatives, in which one plus two cannot be more than three, and where there is no alternative but three." It soaked in at this point that the press had lost the thread of his discourse, so Merino clarified: "And the truth falls in that form before Aristotelian truth, or the classical truth, let's say, that gave certain shadings to the search for it; it has enormous importance in a country like ours that is searching for new paths, looking for new ways to live."

This same admiral justified the government's decision to put him in charge of the economy by saying that he had studied economy as a hobby in courses of the *Encyclopedia Britannica*. And with the same candor he stated on the record his opinion that "War is the most beautiful profession there is. And what is war? The continuation of peace in which all the things peace does not allow are achieved, in order to lead man to the perfect dialectic, which is the extinction of the enemy."

In 1980 when these gems were appearing in the press, I was no longer in Chile. I stayed awhile, but when I felt repression tightening like a noose around my neck, I left. I watched the country and its people change. I tried to adapt and not attract attention, as my grandfather had asked, but it was impossible because in my situation as a journalist I knew too much. At first my fear was something vague and difficult to define, like a bad smell. I discounted the terrible rumors that were circulating, alleging that there was no proof, and when proof was presented to me, I said those were exceptions. I thought I was safe because I wasn't visibly "involved" in politics, in the meantime sheltering desperate fugitives in my home or helping them over embassy walls in search of asylum. I thought that if I were arrested I could explain that I was acting out of humanitarian motives. Apparently I was somewhere on the moon. I broke out in hives from head to foot, I couldn't sleep, and the sound of a car in the street after curfew would leave me trembling for hours. It took me a year and a half to realize the risk I was running, and finally, in 1975, following a particularly agitated and danger-filled week, I left for Venezuela, carrying a handful of Chilean soil from my garden. A month later, my husband and my children joined me in Caracas. I suppose I suffer the affliction of many Chileans who left during that time: I feel guilty for having abandoned my country. I have asked myself a thousand times what would have happened had I stayed, like so many who fought the dictatorship from within, until it was overthrown in 1989. No one can answer that question, but of one thing I am sure: I would not be a writer had I not experienced that exile.

From the instant I crossed the cordillera of the Andes one rainy winter morning, I unconsciously began the process of inventing a country. I have flown over those mountains many times since, and I am always deeply moved because the memory of that morning assaults me full-force as I look down on the magnificent spectacle of the mountains. The infinite solitude of those white peaks, those dizzying abysses, the blue depths of the sky, symbolizes my farewell to Chile. I never imagined I would be gone for so long. Like all Chileans— except the military—I was convinced that given our tradition, the soldiers would soon return to their barracks, there would be a new election, and we would have a democratic government again. I must have intuited something in regard to the future, however, because I spent my first night in Caracas crying inconsolably in a borrowed bed. Deep down, I sensed that something had ended forever, and that my life was taking a new direction. I have felt the pangs of nostalgia ever since that first night, and they did not lessen for many years—until the dictatorship fell and I again stood on the soil of my country. Through the intervening years, I lived with my eyes turned south, listening to the news, waiting for the moment I could go back, as I selected my memories, altered some events, exaggerated or ignored others, refined my emotions, and so gradually constructed the imaginary country in which I have sunk my roots.

> *There are exiles that gnaw and others*
> *that are like consuming fire.*
> *There is heartache for the murdered country*
> *that rises from below*
> *from feet and from roots*
> *and suddenly the man is suffocating,*
> *he no longer knows corn tassels,*
> *the guitar has been silenced,*
> *there is no air for that mouth*
> *he can't live without a land,*
> *and then he falls to his knees*
> *not onto native soil, but into death.*
>
> PABLO NERUDA,
> "EXILES" FROM *CANTOS*
> *CEREMONIALES*

Among the notable changes produced by the values and the economic system instituted by the dictatorship is that ostentation became fashionable. If you aren't wealthy, you should go into debt to look as if you are, even if you have holes in your socks. Consumerism is the current ideology in Chile, as it is in most places in the world. Economic policy, negotiating, and corruption that reached levels never seen in the nation, created a new caste of millionaires. One of the positive effects of it was that it shattered the wall separating the social classes; old family names are no longer the only passport for being accepted in society. Those who thought of themselves as aristocrats were swept off the map by young impresarios and technocrats riding their chrome motorcycles and driving their Mercedes-Benzes, and by a few military officers who got rich in key posts of the government, industry, and banking. For the first time, men in uniform were everywhere: ministries, universities, corporations, salons, and clubs.

The hard question is why at least one third of Chile's total population backed the dictatorship, even though for most life wasn't easy and even adherents of the military government lived in fear. Repression was far reaching, although there's no doubt that the poor and the leftists suffered most. Everyone felt he was being spied on, no one could say that he was completely safe from the claws of the state. It is a fact that information was censored and brainwashing was the goal of a vigorous propaganda machine; it is also true that the opposition lost many years and a lot of blood before it could get organized. But none of this explains the dictator's popularity. The percentage of the population that approved of him was not motivated solely by fear: Chileans like authority. They believed that the military was going to "clean up" the country. "They put an end to delinquency, we don't see walls defaced with graffiti any more, everything is clean, and, thanks to the curfew, our husbands get home early," one friend told me. For her those things compensated for the loss of civil rights because she wasn't directly affected: she was in the fortunate position of not having her children lose their jobs without compensation, or of being arrested. I understand why the economic right, which historically has not been characterized as a defender of democracy and which during those years made more money than ever before, backed the dictatorship, but what about the rest? I haven't found a satisfactory answer to that question, only conjectures.

Pinochet represented the intransigent father, capable of imposing strict discipline. The three years of the Unidad Popular were a time of experimentation, change, and disorder; the country was weary. Repression put an end to politicking, and neoliberalism forced Chileans to work, keep their mouths closed, and be productive, so that corporations could compete favorably in international markets. Nearly everything was privatized, including health, education, and social security. The need to survive drove private initiative. Today Chile not only exports more salmon than Alaska, but also, among hundreds of other nontraditional products, ships out frogs' legs, goose feathers, and smoked garlic. The U.S. press celebrated the triumph of Pinochet's economic system and gave him credit for having turned a poor country into the star of Latin America. None of the indices, however, revealed the distribution of wealth; nothing was known of the poverty and uncertainty in which several million people were living. There was no mention of the soup kitchens in poor neighborhoods that fed thousands of families—there were more than five hundred in Santiago alone—or of the fact that private charities and churches were trying to replace the social services that are the responsibility of the state. There was no open forum for discussing government actions or those of businessmen; public services were handed over to private companies, and foreign corporations acquired natural resources such as forests and oceans, which have been exploited with very little ecological conscience. A callous society was created in which profit is sacred; if you are poor it's your own fault, and if you complain, that makes you a Communist. Freedom consists of having many brand names to choose from when you go out to buy on credit.

The figures of economic growth, which won the *Wall Street Journal*'s praise, did not represent real development, since 10 percent of the population possessed half the nation's wealth and there were a hundred persons who earned more than the state spent on all social services combined. According to the World Bank, Chile is one of the countries with the worst distribution of income, right alongside Kenya and Zimbabwe. The head of a Chilean corporation earns the same or more than his equivalent in the United States, while a Chilean laborer earns approximately fifteen times less than a North American worker. Even today, after more than a decade of democracy, the disparities in wealth are staggering because the economic model hasn't changed. The three presidents who followed

Pinochet have had their hands tied; the right controls the economy, the Congress, and the press. Chile, nonetheless, has proposed to become a developed country within the span of a decade, which is possible if, in fact, wealth is redistributed in a more equitable fashion.

Who was Pinochet, really? This military man who so deeply marked Chile with his capitalist revolution and two decades of repression? (He is still alive but I use the past tense because he is under house arrest and the country is trying to forget he existed. He belongs to the past, even though his shadow still darkens the country.) Why was he so feared? Why was he admired? I never met him personally and I didn't live in Chile during the greater part of his government, so I can only judge him by his actions and what others have written about him. I suppose that to understand Pinochet you need to read novels like Mario Vargas Llosa's *Feast of the Goat* or Gabriel García Márquez's *Autumn of the Patriarch,* because he had a lot in common with the typical figure of the Latin American caudillo so aptly described by those authors. He was a crude, cold, slippery, authoritarian man who had no scruples or sense of loyalty other than to the army as an institution—though not to his comrades in arms, whom he had killed according to his convenience, men like General Carlos Prats and others. He believed he was chosen by God and history to save his country. He was fond of medals and military paraphernalia, to the degree that he established a foundation bearing his name to promote and preserve his image. He was astute and suspicious, but he could be genial, and at times even likeable. Admired by some, despised by others, feared by all, he was possibly the man in our history who has held the greatest power in his hands for the longest period of time.

CHILE IN MY HEART

In Chile people try to avoid talking about the past. The youngest generations believe the world began with them; anything that happened before they were born doesn't interest them. And it may be that the rest of the population shares a collective shame regarding what took place during the dictatorship, the same feeling that Germany had after Hitler. Both young and old want to avoid discord. No one wants to be led into discussions that drive even deeper wedges. Furthermore, people are too busy trying to get to the end of the month with a salary that doesn't stretch far enough, and quietly doing their job so they won't be fired, to be concerned about politics. It's assumed that digging too much into the past can "destabilize" the democracy and provoke the military, a fear that is totally unfounded, since the democracy has been strengthened in recent years—since 1989—and the military has lost prestige. Besides, this is not a good time for military coups. Despite its many problems—poverty, inequality, crime, drugs, guerrilla wars—Latin America has opted for democracy, and for its part, the United States is beginning to realize that its policy of supporting tyranny does not solve problems—it merely creates new ones.

The military coup didn't come out of nowhere; the forces that upheld the dictatorship were there, we just hadn't perceived them. Defects that had lain there beneath the surface blossomed in all their glory and majesty during that period. It isn't possible that repression on such a grand scale could have been organized overnight unless a totalitarian tendency already existed in a sector of the society; apparently we were not as democratic as we believed. As for the government of Salvador Allende, it wasn't as innocent as I like to imagine; it suffered from ineptitude, corruption, and pride. In real life, it may not always be easy to distinguish between heroes and villains, but I can assure you that in democratic governments, including that of the Unidad Popular, there was never the cruelty the nation has suffered every time the military intervenes.

Like thousands of other Chilean families, Miguel and I left with our two children because we didn't want to go on living in a dictatorship. That was 1975. The country we chose to emigrate to was Venezuela because it was one of the last remaining democracies in Latin America, shaken by military coups but one of the few countries that would grant us visas and the opportunity for work. Neruda says:

How can I live so far away
from what I loved, what I love?
From the changing seasons, clothed
in steam and cold smoke?

(Strangely enough, the thing I missed the most during those years of self-imposed exile were the seasons of the year. In the eternal green of the tropics, I was a complete stranger.)

In the seventies Venezuela was experiencing the peak of the oil boom, black gold gushed from its soil like a raging river. Everything seemed easy; with a minimum of work and decent connections, people lived better than anywhere else. Money flowed like water, and people spent it as if there were no tomorrow. More champagne was consumed in Venezuela than in any other country in the world. For those of us who had gone through the economic crisis of the government of the Unidad Popular, in which toilet paper was a luxury, and then escaped tremendous repression, Venezuela was beyond our comprehension. We couldn't take in the leisure time, the easy money, and the freedom of that country. We Chileans, so serious, so sober and prudent, so fond of rules and legalisms, couldn't understand that unfettered joy and indifference to discipline. Accustomed to euphemisms, we were offended by the frankness of speech. We Chileans numbered several thousand, and soon we were joined by others escaping from the "dirty wars" in Argentina and Uruguay. Some arrived with marks of recent imprisonment; all came with an air of defeat.

My husband found work in the interior of the country and I stayed in Caracas with our two children, who begged me every day to go back to Chile, where they had left grandparents, friends, school—in short, everything they knew. That separation from my husband proved fatal; I believe it marked the beginning of the end of our lives as man and wife. We weren't the exception, because most of the couples who left Chile together ended up separated. Far away from country and family, the pairs found themselves face to face, naked and vulnerable, without the family pressure, the social crutches and routines that hold two people together. The circumstances were no help: fatigue, fear, insecurity, poverty, confusion; if in addition you were separated geographically, as happened with us, the prognosis was poor. Unless you're lucky and your bond is very strong, love dies.

I couldn't find a job as a journalist. What I'd done earlier in Chile made little impression, partly because exiles tend to inflate their credentials and in the end no one believes much of anything; there were false doctors who had barely graduated from high school and real doctors who ended up driving taxis. I didn't know a soul, and there, as in the rest of Latin America, you don't get anywhere without connections. I had to earn a living by taking insignificant jobs, none of which is worth mentioning. I didn't understand the Venezuelan temperament, I confused their deeply felt sense of equality with bad manners, their extroversion with pedantry, their emotionalism with immaturity. I came from a country in which violence had been institutionalized and yet I was shocked by how quickly Venezuelans lost control. (Once at a movie theater a woman pulled a pistol from her handbag because I accidentally sat in a seat she had reserved.) I didn't know their customs; for example, they rarely say no because they think it's rude: they would rather say "Come back tomorrow." I would go to look for a job and they would interview me with a great show of friendliness, offer me coffee, and say good-bye with a firm handshake and that "Come back tomorrow." So I would come back the next day, and the same routine would be repeated until finally I gave up. I felt that my life was a failure; I was thirty-five years old and I thought I had no future before me except to grow old and die of boredom. Now when I remember that time, I realize that opportunities existed but I didn't see them; I was confused and fearful, and incapable of dancing to their tune. Instead of making an effort to learn about the land that had so generously taken me in, and learn to love it, I was obsessed with going home to Chile. When I compare my experience as an exile with my current situation as an immigrant, I can see how different my state of mind is. In the former instance, you are forced to leave, whether you're escaping or expelled, and you feel like a victim who has lost half her life; in the latter it's your own decision, you are moving toward an adventure, master of your fate. The exile looks toward the past, licking his wounds, the immigrant looks toward the future, ready to take advantage of the opportunities within his reach.

We Chileans in Caracas got together to listen to the records of Violeta Parra and Víctor Jara, to exchange posters of Allende and Che Guevara, and to repeat a thousand times over the same rumors about our distant homeland. Every time we met we ate empanadas; I got so sick of them that to this day I can't eat one. Every day new compatriots arrived with terrible stories, swearing that the dictatorship was about to collapse, but months went by and, far from collapsing, that government seemed stronger and stronger, despite internal protests and an enormous groundswell of international solidarity. Now no one confused Chile with China, and no one asked why we didn't wear pineapples on our heads; the figure of Salvador Allende and the resulting political events had placed the country on the map. One photograph that made the rounds became famous: the military junta with Pinochet in the center, arms crossed, dark glasses, protruding bulldog chin—a true cliché of Latin American tyranny. Strict censorship of the press prevented most Chileans from realizing that such solidarity existed outside the country. I had lived a year and a half under that censorship, and I didn't know that elsewhere the name of Allende had become a symbol, and when I left the country I was amazed at the reverential respect my surname occasioned. Unfortunately, that consideration didn't help me find work, which I desperately needed.

From Caracas I wrote to my grandfather, whom I hadn't had the courage to say good-bye to; I wouldn't have been able to explain my reasons for escaping without admitting that I had disobeyed his instructions not to get into trouble. In my letters I painted a rosy picture of our lives but it didn't take enormous perception to read between the lines, and my grandfather must have guessed my true situation. Soon that correspondence turned into pure nostalgia, a patient exercise of remembering the past and the land I had left behind. I started reading Neruda again, and quoted him in my letters. Sometimes my grandfather answered with lines from other, older, poets.

I won't repeat here the details of those years, the good things that happened, and the bad, such as failed love affairs, loneliness, struggles, and sorrows, because I have already told about them elsewhere. It's enough to say that the feelings of loneliness and of being an outsider that I'd had since I was a child were accentuated. I was cut off from reality, submerged in an imaginary world, while right before my eyes my children were growing up and my marriage was falling apart. I tried to write, but all I could do was go over and over the same ideas. At night, after my family went to bed, I locked myself in the kitchen, where I spent hours pounding the keys of an old Underwood, filling pages and pages with the same sentences; afterward I would rip them to pieces, like Jack Nicholson in that hair-raising film *The Shining,* which left half the world with nightmares for months. Nothing remains of those efforts . . . nothing but confetti. And so seven years went by.

On January 8, 1981, I began another letter to my grandfather, who by then was nearly a hundred and was dying. From the first sentence, I knew it wasn't a letter like the others and that it might never reach the hands of the person to whom I was writing. I wrote to ease my anguish, because that old man, the storehouse of my oldest memories, was ready to leave this world. Without him, the anchor in the land of my childhood, my exile seemed definitive. Naturally I wrote about Chile and my far-flung family. I had more than enough material to write about with the hundreds of stories that had poured from his lips over the years: our proto-macho forefathers; my grandmother, who moved the sugar bowl with pure spiritual energy; Aunt Rosa, who died at the end of the nineteenth century, and whose ghost appeared at night to play the piano; the uncle who tried to cross the cordillera in a dirigible; and all those other characters who shouldn't simply fade into oblivion. When I told those tales to my children, they looked at me with pitying expressions and rolled their eyes. After crying so hard to go back, Paula and Nicolás had finally adapted to Venezuela and didn't want to hear anything about Chile, and especially not their bizarre relatives. They never took part in the nostalgic conversations among us older exiles, in the failed attempts to make Chilean dishes with Caribbean ingredients, or in the pathetic celebrations of national holidays we improvised in Venezuela. My children were embarrassed to be foreigners.

Soon I lost track of where that strange letter was going, but I kept writing it for a whole year, at the end of which my grandfather had died and my first novel was sitting on the kitchen table: *The House of the Spirits*. If someone had asked what it was about, I would have said that it was an attempt to recapture my lost country, to reunite my scattered family, to revive the dead and preserve their memories, which were beginning to be blown away in the whirlwind of exile. It wasn't a small thing I was attempting. . . . Now I have a simpler explanation: I was dying to tell that story.

I have a romantic image of a Chile frozen at the beginning of the seventies. For years I believed that when democracy was restored everything would be as it had been before, but even that frozen image was deceptive. Maybe the place I'm homesick for never existed. Now when I visit, I must compare the real Chile to the sentimental image I've carried for twenty-five years. Since I've lived outside the country for so long, I tend to exaggerate the virtues of our national character and forget the disagreeable aspects. I forget the snobbishness and hypocrisy of the upper class; I forget how conservative and macho the greater part of the society is; I forget the crushing authority of the Catholic Church. I am frightened by the rancor and violence nourished by inequality, but I am also moved by the good things that have survived despite all that has happened, such as the immediate familiarity of our relationships, the affectionate way we greet one another with kisses, the twisted sense of humor that always makes me laugh, the friendship, hope, simplicity, and congeniality, the solidarity in difficult times, the sympathy, the indomitable courage of mothers, the patience of the poor.

I have constructed an idea of my country the way you fit together a jigsaw puzzle, by selecting pieces that fit my design and ignoring the others. My Chile is poetic and poor, which is why I discard the evidence of a modern, materialistic society in which a person's value is measured by wealth, fairly acquired or otherwise, and insist on seeing signs everywhere of my country of old. I have also created a version of myself that has no nationality, or, more accurately, many nationalities. I don't belong to one land, but to several, or perhaps only to the ambit of the fiction I write. I can't pretend to know what part of my memory is reliable and how much I've invented, because the job of defining the line between them is beyond my ability. My granddaughter Andrea wrote a composition for school in which she said that she liked her "grandmother's imagination." I asked her what she was referring to, and without hesitation she replied, "You remember things that never happened." Don't we all do that? I have read that the mental process of imagining and that of remembering are so much alike that they are nearly indistinguishable. Who can define reality? Isn't everything subjective? If you and I witness the same event, we will recall it and recount it differently. Comparing the versions of our childhood that my brothers tell, it's as if each of us had been on a different planet. Memory is conditioned by emotion; we remember better, and more fully, things that move us, such as the joy of a birth, the pleasure of a night of love, the pain of a loved one's death, the trauma of a wound. When we call up the past, we choose intense moments— good or bad—and omit the enormous gray area of daily life.

If I had never traveled, if I had stayed on, safe and secure in the bosom of my family, if I had accepted my grandfather's vision and his rules, it would have been impossible for me to recreate or embellish my own existence, because it would have been defined by others and I would merely be one link more in a long family chain. Moving about has forced me, time after time, to readjust my story, and I have done that in a daze, almost without noticing, because I have been too preoccupied with the task of surviving. Most of our lives are similar, and can be told in the tone used to read the telephone directory—unless we decide to give it a little oomph, a little color. In my case, I have tried to polish the details and create my private legend, so that when I am in a nursing home awaiting death I will have something to entertain the other senile old folks with.

I wrote my first book by letting my fingers run over the typewriter keys, just as I am writing this, without a plan. I needed very little research because I had it all inside, not in my head but in that place in my chest where I felt a perpetual knot. I told about Santiago in the time of my grandfather's youth, just as if I'd been born then; I knew exactly how a gas lamp was lit before electricity was installed in the city, just as I knew the fate of hundreds of prisoners in Chile during that same period. I wrote in a trance, as if someone was dictating to me, and I have always attributed that favor to the ghost of my grandmother, who was whispering into my ear. Only one other time have I been gifted with a book dictated from that other dimension, and that was when I wrote my memoir *Paula* in 1993. I have no doubt that in writing that book I received help from the benign spirit of my daughter. Who, really, are these and the other spirits who live with me? I haven't seen them floating around the hallways of my home, wrapped in white sheets, nothing as interesting as that. They are simply memories that come to me and that from being caressed so often gradually acquire flesh. That happens with people, and also with Chile, that mythic country that from being missed so profoundly has replaced the real country. That country inside my head, as my grandchildren describe it, is a stage on which I place and remove objects, characters, and situations at my whim. Only the landscape remains true and immutable; I am not a foreigner to the majestic landscape of Chile. My tendency to transform reality, to invent memory, disturbs me, I have no idea how far it may lead me. Does the same thing happen with people? If, for example, I saw my grandparents or my daughter for an instant, would I recognize them? Probably not, because in looking so hard for a way to keep them alive, remembering them in the most minimal details, I have been changing them, adorning them with qualities they may not have had. I have given them a destiny much more complex than the ones they lived. In any case, I have been very lucky because that letter to my dying grandfather saved me from desperation. Thanks to it, I found a voice and a way to overcome oblivion, which is the curse of vagabonds like me. Before me opened the road-of-no-return of literature, which I have stumbled down the last twenty years, and which I hope to follow as long as my patient readers will put up with me.

Although that first novel gave me a fictitious country, I never stopped loving the other one, the one I had left behind. The military government was solidly entrenched in Chile, and Pinochet was ruling with absolute power. The economic policy of the Chicago Boys, as Milton Friedman's disciples were known, had been imposed by force; it could not have been done any other way. Entrepreneurs were enjoying enormous privileges, while workers had lost most of their rights. Those of us who had left thought that the dictatorship would remain in power for some time, but in truth a valiant opposition was growing inside the country, one that finally would lead to restoring the toppled democracy. In order to do that it would be necessary to set aside the many party squabbles and join together in the Concertación coalition . . . but that would be seven years later. In 1981 few could imagine that possibility.

Up until then my life in Caracas, where we had lived for ten years, had gone by in complete anonymity, but books attract a little bit of attention. Finally I resigned from the school where I was working and dived into the uncertainty of literature. I had another novel in mind, this one situated somewhere in the Caribbean; I thought I was through with Chile and that it was time to write about the land that gradually was becoming my adopted country. Before I began writing *Eva Luna,* I had to do a lot of research. To describe the odor of a mango or shape of a palm, I had to go to the market to smell the fruit and to the plaza to look at the trees, which hadn't been the case with a Chilean peach or willow tree. I have Chile so deep inside me that I think I know it backward and forward, but when I write about a different place, I have to study it.

In Venezuela, a splendid land of assertive men and beautiful women, I was liberated at last from the discipline of English schools, the rigor of my grandfather, Chilean modesty, and the last vestiges of that formality in which, the good daughter of diplomats, I had been brought up. For the first time I felt comfortable in my body and stopped worrying about what others thought of me. In the meantime my marriage had deteriorated beyond repair, and once our children left the nest to go to the university there was no further reason to stay together. My husband and I were amicably divorced. We were so relieved by this decision that as we said good-bye we bowed reverential Japanese bows for several minutes. I was forty-five years old, but I didn't look bad for my age—at least that's what I thought until my mother, always an optimist, warned me that I was going to spend the rest of my life alone. Nevertheless, three months later, during a long promotion tour in the United States, I met William Gordon, the man who was written in my destiny, as my clairvoyant grandmother would say.

THIS COUNTRY INSIDE MY HEAD

B efore you ask me why a leftist with my surname chose to live in the Yankee empire, I will tell you that it wasn't by plan, not by any stretch of the imagination. Like almost all the major milestones in my life, it happened by chance. If Willie had been in New Guinea, most probably I would be there now, dressed in feathers. I suppose there are people who do plan their lives, but I stopped doing that a long time ago because my blueprints never get used. About every ten years I take a look back and can see the map of my journey—well, that is if it can be called a map, it looks more like a plateful of noodles. If you live long enough to review the past, it's obvious that all we do is walk in circles. The idea of settling in the United States never crossed my mind; I believed that the CIA had incited the military coup in Chile for the sole purpose of ruining my life. Over the years I have become more modest. I had only one reason to become just one more among the millions of immigrants pursuing the American Dream: lust at first sight.

Willie had two divorces behind him and a string of affairs that he can barely remember. He had been single for eight years, his life was a disaster, and he was still waiting for the tall blonde of his dreams when I came along. He had barely looked down and separated me from the design on the carpet, when I informed him that in my youth I had been a tall blonde; that was what caught his attention. What about him attracted me? I could tell that he was a strong person, the kind who may fall to his knees but who gets right back up on his feet. He was different from the average Chilean: he didn't complain, he didn't blame others for his problems, he accepted his karma, he wasn't looking for a mother, and it was obvious he didn't need a geisha to bring him breakfast in bed or to lay out his clothes for the next day. I could see that he didn't belong to the school of the stoics, like my grandfather, it was too obvious that he enjoyed his life, but he did have the same stoic stability. Besides that, he'd traveled a lot; which is always seductive to us Chileans, who are basically insular people. At twenty he'd gone around the world, hitchhiking and sleeping in cemeteries. (He explained to me that they're very safe, no one goes there at night.) He had been exposed to different cultures, he was broad-minded, and he was tolerant and curious. He also spoke good Spanish—with the accent of a Mexican bandit—and he had tattoos. In Chile, only criminals sport tattoos, so I thought he was really sexy. He could order dinner in French, Italian, and Portuguese, and he knew how to mumble a few words in Russian, Tagalog, Japanese, Mandarin, Swahili, and Farsi. Years later I discovered that he invented them, but by then it was too late. To top it all off, he could speak English as well as any North American manages to master the language of Shakespeare.

We found a way to be together for two days, and then I had to continue my tour, but at the end I decided to return to San Francisco for a week to see whether I could get him out of my head or whether lust had turned into love. This is a very Chilean way to behave; any of my female compatriots would have done the same. We Chilean women are ferociously decisive in two things: in defending our cubs and in trapping a man. We have a strongly developed nesting instinct; an adventure isn't enough for us, we want to form a household and if possible, have children. Imagine! When I arrived at his house, uninvited, Willie, in a panic, tried to make his escape, but he wasn't really a serious opponent for me. I took one running leap and was on him like a prizefighter. Finally he agreed, gritting his teeth, that I was the closest thing to a tall blonde he was ever going to get, and we got married. That was 1987.

To be near Willie, I was ready to give up a lot, but not my

children or my writing, so as soon as I got my residence papers I began the process of moving Paula and Nicolás to California. I had quickly become enamored of San Francisco, a happy, tolerant, open, and cosmopolitan city—and so different from Santiago! My new home was founded by adventurers, prostitutes, merchants, and preachers, all of whom flocked there in 1849, drawn by the Gold Rush. I wanted to write about that intriguing period of greed, violence, heroism, and conquest, perfect material for a novel. In the mid-nineteenth century the surest route to California from the east coast of the United States, or from Europe, went right by Chile. Ships had to sail through the Strait of Magellan or around Cape Horn. Those were dangerous odysseys, but worse was crossing the North American continent in a wagon or slogging through the malaria-infected jungles of the Isthmus of Panama. Chileans learned of the discovery of gold before the news spread in the United States, and they came en masse: they had a long tradition of mining and they liked an adventure. We have a name for our compulsion for following where a road leads, we say that we're *patiperros,* because we roam like yappy little strays sniffing a trail, with no fixed direction. We need escape, but as soon as we cross the cordillera we begin to miss home, and always come back. We're good travelers and terrible emigrants: nostalgia is always nipping at our heels.

Willie's family, and Willie's life, were chaotic, but instead of running away, as any reasonable person would do, I rushed him "straight on, in the name of Chile," like the war cry of the soldiers who took the Arica promontory in the nineteenth century. I was determined to win my place in California and in the heart of that man, cost what it may. In the United States, everyone, with the exception of the Indians, descends from someone who came from somewhere else; there was nothing special about my case. The twentieth century was the century of immigrants and refugees; the world had never seen so many humans fleeing violence or poverty abandon their place of origin to start a new life in a new land. My family and I are part of that diaspora; it isn't as bad as it sounds. I knew that I would never assimilate completely, I was too old to melt in that famous Yankee pot. I look like a Chilean, I dream, cook, make love, and write in Spanish, and most of my books have a pronounced Latin American flavor. I was convinced that I would never be a Californian, but I wouldn't pretend to be one either; all I aspired to was to earn a driver's license and learn enough English to order food in a restaurant. I didn't dream I would get much more.

I've had to work several years to adapt to California, but the process has been entertaining. Writing a book about Willie's life, *The Infinite Plan,* helped a lot because it forced me to travel across the state and study its history. I remember how offended I was at first by the gringos' direct manner of speaking—until I realized that most of them are considerate and courteous. I couldn't believe what hedonists they were, until I caught the fever and ended up soaking in a Jacuzzi surrounded by aromatic candles (meanwhile

my grandfather is whirling in his grave at such wantonness).
I've been so thoroughly incorporated into the California
culture that I practice meditation and go to a therapist, even
though I always set a trap: during my meditation I invent
stories to keep from being bored, and in therapy I invent
others to keep from boring the psychologist. I have adapted
to the rhythm of this extraordinary place; I have favorite
spots where I spend time leafing through books and walking and talking with friends; I like my routines, the seasons
of the years, the huge oaks around my house, the scent of
my cup of tea, the long nocturnal lament of the siren that
warns ships of fog in the bay. I eagerly await the
Thanksgiving turkey and the kitschy splendor of Christmas.
I even take part in the obligatory Fourth of July picnic. And
by the way, that picnic, like everything else in this land, is a
model of efficiency: you drive at top speed, set up in a pre-
viously reserved space, spread out the baskets, bolt your
food, kick the ball, and rush home to avoid the traffic. In
Chile, a similar project would take three days.

The North Americans' sense of time is very special. They
are short on patience. Everything must be quick, including
food and sex, which the rest of the world treats ceremoni-
ously. Gringos invented two terms that are untranslatable into
most languages: "snack" and "quickie," to refer to eating
standing up and loving on the run . . . that, too, sometimes
standing up. The most popular books are manuals: how to
become a millionaire in ten easy lessons, how to lose fifteen
pounds a week, how to recover from your divorce, and so on.
People always go around looking for shortcuts and ways to

escape anything they consider unpleasant: ugliness, old age, weight, illness, poverty, and failure in any of its aspects.

This country's fascination with violence never ceases to shock me. It can be said that I have lived in interesting circumstances, I've seen revolutions, war, and urban crime, not to mention the brutalities of the military coup in Chile. Our home in Caracas was broken into seventeen times; almost everything we had was stolen, from a can opener to three cars, two from the street and the third after the thieves completely ripped off our garage door. At least none of them had bad intentions; one even left a note of thanks stuck to the refrigerator door. Compared to other places on earth, where a child can step on a mine on his way to school and lose two legs, the United States is safe as a convent, but the culture is addicted to violence. Proof of that is to be found in its sports, its games, its art, and, certainly not least, its films, which are bloodcurdling. North Americans don't want violence in their lives, but they need to experience it indirectly. They are enchanted by war, as long as it's not on their turf.

The racism, on the other hand, didn't shock me, even though according to Willie it is the most serious problem in the country, because for forty-five years I had experienced the class system in Latin America, where the poor and mestizos— African or Indian—live in ineradicable segregation, as if it were the most natural thing in the world. At least in the United States there is an awareness of the conflict, and most North Americans, most of the time, fight against racism.

When Willie visits Chile, he is an object of curiosity to

my friends and to children in the street because of his undeniably foreign looks, which he accentuates by wearing an Aussie hat and cowboy boots. He likes my country, he says it's like California was forty years ago, but he feels out of place there, the way I do in the United States. I understand the language but I don't know the codes. When we get together with friends, I can't really participate in the conversation, because I don't know the events or the people they're talking about, I didn't see the same movies when I was young, I didn't dance to the epileptic guitar of Elvis, I didn't smoke marijuana or protest against the war in Vietnam. I can't even follow the political jokes, because I see very little difference between Democrats and Republicans. And I'm a real foreigner for not sharing the national fascination with President Clinton's amorous dalliance; after I saw Miss Lewinsky's drawers for the fourteenth time on TV, I lost interest. Even baseball is a mystery to me, I can't understand such passion for a group of heavyset men waiting for a ball that never comes. In California I'm a misfit; I wear silk while the rest of the population wears sneakers, and I order beef when everyone else is on a kick for tofu and green tea.

The thing I most appreciate about my situation as an immigrant is the marvelous sense of freedom. I come from a very traditional culture, from a closed society, where each of us carries from birth the karma of his ancestors and where we constantly feel watched and judged. A stain on one's honor cannot be cleansed. A child who steals crayons in kindergarten is branded as a thief for the rest of his life.

In the United States, in contrast, the past doesn't matter; no one asks your last name; the son of a murderer can be president . . . as long as he's white. You can make mistakes because new opportunities abound, you just move to a new state and change your name and start a new life. Spaces are so vast that roads never end.

At first Willie, condemned to live with me, felt as uncomfortable with my Chilean ideas and customs as I felt with him. We had major problems, among them that I tried to impose my antiquated norms of family life upon his children and that he had no sense of romanticism. We also had minor problems, such as my being incapable of working the household appliances, and his snoring, but gradually we have overcome these differences. Maybe that's what marriage is about, nothing more than that: being flexible. As an immigrant I have tried to preserve the Chilean virtues that I like and to renounce the prejudices that are as confining as a straitjacket. I have accepted this country. To love a place you must participate in the community and give back something in return for all you receive. I believe I have done that. There are many things I admire about the United States and others I would like to change, but isn't that always true? A country, like a husband, is always open to improvement.

One year after I moved to California, in 1988, the situation changed in Chile; Pinochet had lost the referendum and the country was ready to reinstate democracy. So I went back. I went with fear; I didn't know what I was going to find, and I nearly didn't recognize Santiago or its people: everything was different. The city was filled with gardens and modern buildings, seething with traffic and commerce, energetic and fast-paced and progressive. But there were feudal backwashes, such as maids in blue aprons taking their elderly charges in the wealthy barrios for walks, and beggars at every stoplight. Chileans were cautious; they respected hierarchies and dressed very conservatively—men in ties, women in skirts—and in many government offices and private enterprises, employees were wearing uniforms, like flight attendants. I realized that many of the people who had stayed and suffered in Chile considered those of us who left to be traitors, and believed that life had been much easier for us. There were many exiles, on the other hand, who accused those who stayed in the country of collaborating with the dictatorship.

The candidate of the Concertación Party, Patricio Alwyn, had won by a narrow margin; the presence of the military was still intimidating, and people were quiet and frightened as they went about their lives. The press was still censored; the journalists who interviewed me, trained in discretion, asked careful, ingenuous questions, and then didn't publish the answers. The dictatorship had done everything possible to erase recent history and the name of Salvador Allende. On the return flight, when I saw San Francisco Bay from the air, I gave a sigh of exhaustion and, without thinking, said: Back home at last. It was the first time since I'd left Chile in 1975 that I felt I was "home."

I don't know whether my home is the place where I

live or simply Willie. We have been together a number of years, and it seems to me that he is the one territory I belong in, where I'm not a foreigner. Together we have survived many ups and downs, great successes and great losses. The most profound sorrow has come from the tragedies of our daughters. In the space of one year, Jennifer died of an overdose and Paula of a rare genetic condition called porphyria, which caused her to sink into a long coma and finally took her life. Willie and I are strong and stubborn, and it was difficult for us to admit that our hearts were broken. It took time and therapy before we could finally put our arms around one another and weep together. The mourning was a long voyage through hell, from which I was able to emerge with his help and that of my writing.

In 1994 I went back to Chile, looking for inspiration, a trip I have since repeated yearly. I found my compatriots more relaxed and the democracy stronger, although conditioned by the presence of a still-powerful military and by the senators Pinochet had appointed for life in order to control the Congress. The government had to maintain a delicate balance among the political and social forces. I went to working-class neighborhoods where people had once been contentious and organized. The progressive priests and nuns who had lived among the poor all those years told me that the poverty was the same but that the solidarity had disappeared, and that now crime and drugs, which had become the most serious problem among the young—had been added to the issues of alcoholism, domestic violence, and unemployment.

The rules to live by were: try to forget the past, work for the future, and don't provoke the military for any reason. Compared to the rest of Latin America, Chile was living in a good moment of political and economic stability; even so, five million people were still below the poverty level. Except for the victims of repression, their families, and a few organizations that kept a watch out for civil rights violations, no one spoke the words *disappeared* or *torture* aloud. That situation changed when Pinochet was arrested in London, where he had gone for a medical check-up and to collect his commission for an arms deal. A Spanish judge charged him with murdering Spanish citizens, and requested his extradition from England to Spain. The general, who still counted on the unconditional support of the armed forces, had for twenty-five years been isolated by the adulators who always congregate around power. He had been warned of the risks of travel abroad, but he went anyway, confident of his impunity. His surprise at being arrested by the British can be compared only to that of everyone in Chile, long accustomed to the idea that he was untouchable. By chance, I was in Santiago when that occurred, and I witnessed how within the course of a week a Pandora's box was opened and all the things that had been hidden beneath layers and layers of silence began to emerge. In those first days there were turbulent street demonstrations by Pinochet's supporters, who threatened nothing less than a declaration of war against England or a commando raid to rescue the prisoner. The nation's press, frightened, wrote of the insult to the Esteemed Senator-for-Life and to the honor and sovereignty of the nation, but a week later demonstrations in his support had become minimal, the military were keeping mute, and the tone had changed in the media: now they referred to the "ex–dictator, arrested in London." No one believed that the English would hand over the prisoner to be tried in Spain, which in fact didn't happen, but in Chile the fear that was still in the air diminished rapidly. The military lost prestige and power in a matter of days. The tacit agreement to bury the truth was over, thanks to the actions of that Spanish judge.

On that trip I traveled through the south. Again I lost myself in the prodigious nature of my country and met with faithful friends to whom I am closer than to my brothers; in Chile, friendship is forever. I returned to California renewed and ready to work. I assigned myself a subject as far removed from death as possible and wrote *Aphrodite,* some ramblings about gluttony and lust, the only cardinal sins worth paying a penance for. I bought a ton of cookbooks and quite a few about eroticism, and I made excursions to the gay district of San Francisco, where for several weeks I scavenged through the pornography shops. (That kind of investigation would have been difficult in Chile. On the off chance that such material existed, I would never have dared buy it: it would have placed my family's honor in jeopardy.) I learned a lot. It's a shame that I acquired this knowledge so late in my life, when I don't have anyone to practice with: Willie made it clear that he is not disposed to hanging a trapeze from the ceiling.

That book helped me emerge from the depression I had sunk into with the death of my daughter. Since that time I have written a book a year. I'm never short on ideas, only time. With Chile and California in mind, I wrote *Daughter of Fortune* and *Portrait in Sepia,* books in which the characters travel back and forth between my two countries.

In conclusion I want to add that the United States has treated me very well. It has allowed me to be myself, or any version of self it has occurred to me to create. The entire world passes through San Francisco, each person carrying his or her cargo of memories and hopes. This city is filled with foreigners; I am not an exception. In the streets you hear a thousand tongues, temples are raised for all denominations, and the scent of food from the most remote points of the world fills the air. Few people are born here, most are strangers in paradise, as I am. It doesn't matter to anyone who I am or what I do; no one watches me or judges me, they leave me in peace. The negative side of that is that if I drop dead in the street, no one will notice but, in the end, that is a cheap price to pay for liberty. The price I would pay in Chile would be high indeed, because there diversity is not as yet appreciated. In California the only thing that isn't tolerated is intolerance.

My grandson Alejandro's observation about the three years I have left to live forces me to ask myself whether I want to live them in the United States or return to Chile. I don't know the answer. Frankly, I doubt that I would leave my house in California. I visit Chile once or twice a year, and when I arrive a lot of people seem happy to see me, though I think they're even happier when I leave—including my mother, who lives in fear that her daughter will do something foolish, for example, appear on television talking about abortion. I feel great for a few days, but after two or three weeks I begin to miss tofu and green tea.

This book has helped me understand that I am not obligated to make a decision: I can have one foot in Chile and another here, that's why we have planes, and I am not among those who are afraid to fly because of terrorism. I have a fatalistic attitude: no one dies one minute before or one minute after the prescribed time. For the moment California is my home and Chile is the land of my nostalgia. My heart isn't divided, it has merely grown larger. I can live and write anywhere. Every book contributes to the completion of that "country inside my head," as my grandchildren call it. In the slow practice of writing, I have fought with my demons and obsessions, I have explored the corners of memory, I have dredged up stories and people from oblivion, I have stolen others' lives, and from all this raw material I have constructed a land that I call my country. That is where I come from.

I hope that this long commentary answers that stranger's question about nostalgia. Don't believe everything I say: I tend to exaggerate and, as I warned at the beginning, I can't be objective where Chile is concerned. Let's just say, to be completely honest, that I can't be objective, period. In any case, what's most important doesn't appear in my biography or my books, it happens in a nearly imperceptible way in the secret chambers of the heart. I am a writer because I was born with a good ear for stories, and I was lucky enough to have an eccentric family and the destiny of a wanderer. The profession of literature has defined me. Word by word I have created the person I am and the invented country in which I live.

ACKNOWLEDGMENTS

This book is based in memory, but I have been aided by the comments of friends: Delia Vergara, Malú Sierra, Vittorio Cintolessi, Josefina Rosetti, Agustín Huneeus, Cristián Toloza and others. I have also called on works by Alonso de Ercilla y Zúñiga, Eduardo Blanco Amor, Benjamín Subercaseaux, Leopoldo Castedo, Pablo Neruda, Alfredo Jocelyn-Holt, Jorge Larraín, Luis Alejandro Salinas, María Luisa Cordero, Pablo Huneeus, and many more. I am grateful, as always, to my mother, Francisca Llona, and to my stepfather, Ramón Huidobro, for helping me with various dates and for correcting the final text. Also to my faithful agents, Carmen Balcells and Gloria Gutiérrez, to my copy-editor, Jorge Manzanilla, and to my North American editor, Terry Karten.